Incorporating the Digital Commons:
Corporate Involvement in Free and Open Source Software

Benjamin J. Birkinbine

To Kingking !
I bled for this book.

Ben

Critical, Digital and Social Media Studies

Series Editor: Christian Fuchs

The peer-reviewed book series edited by Christian Fuchs publishes books that critically study the role of the internet and digital and social media in society. Titles analyse how power structures, digital capitalism, ideology and social struggles shape and are shaped by digital and social media. They use and develop critical theory discussing the political relevance and implications of studied topics. The series is a theoretical forum for internet and social media research for books using methods and theories that challenge digital positivism; it also seeks to explore digital media ethics grounded in critical social theories and philosophy.

Editorial Board

Published

Critical Theory of Communication: New Readings of Lukács, Adorno, Marcuse, Honneth and Habermas in the Age of the Internet
Christian Fuchs
https://doi.org/10.16997/book1

Knowledge in the Age of Digital Capitalism: An Introduction to Cognitive Materialism
Mariano Zukerfeld
https://doi.org/10.16997/book3

Politicizing Digital Space: Theory, the Internet, and Renewing Democracy
Trevor Garrison Smith
https://doi.org/10.16997/book5

Capital, State, Empire: The New American Way of Digital Warfare
Scott Timcke
https://doi.org/10.16997/book6

The Spectacle 2.0: Reading Debord in the Context of Digital Capitalism
Edited by *Marco Briziarelli and Emiliana Armano*
https://doi.org/10.16997/book11

The Big Data Agenda: Data Ethics and Critical Data Studies
Annika Richterich
https://doi.org/10.16997/book14

Incorporating the Digital Commons: Corporate Involvement in Free and Open Source Software

Benjamin J. Birkinbine

UNIVERSITY OF WESTMINSTER PRESS

University of Westminster Press
www.uwestminsterpress.co.uk

Published by
University of Westminster Press
115 New Cavendish Street
London W1W 6UW
www.uwestminsterpress.co.uk

Cover design: www.ketchup-productions.co.uk
Series cover concept: Mina Bach (minabach.co.uk)

Print and digital versions typeset by Siliconchips Services Ltd.

ISBN (Paperback): 978-1-912656-42-4
ISBN (PDF): 978-1-912656-43-1
ISBN (EPUB): 978-1-912656-44-8
ISBN (Kindle): 978-1-912656-45-5

DOI: https://doi.org/10.16997/book39

The full text of this book has been peer-reviewed to ensure high academic standards. For full review policies, see: http://www.uwestminsterpress.co.uk/site/publish. Competing interests: The author has no competing interests to declare.

Contents

List of Figures

List of Tables

Acknowledgments

This book represents the culmination of more than six years of studying and using free and open source software, which began during my doctoral studies in the School of Journalism and Communication at the University of Oregon. I was first introduced to open source software during the summer of 2011 when a colleague, Jeremy Swartz, took me on a guided tour of the Open Source Convention (OSCON) in Portland, Oregon. At the convention, I was offered three different jobs by companies who had exhibits there. I was told, 'we are looking for people like you'. I was incredibly surprised by this because I knew practically nothing about free and open source software at that time. I had heard of Linux, but that was about it. Nonetheless, all of the exhibitors – including Facebook, Google, the New York Times, and the United States Government – seemed keenly interested in attracting free and open source software programmers to their organisations. That experience piqued my interest in why companies were so interested in open source, and it became the focus of my studies.

Along the way, I have had the privilege of discussing the subject with incredibly smart and patient people whohave helped me. I thank my dissertation committee for their patience, comments, and critiques throughout the process. Janet Wasko was an incredibly gracious and accommodating adviser throughout the process. Thanks for the support during my time at Oregon, the one-on-one meetings and, of course, all the fun. Bish Sen provided critical feedback throughout the process and always pushed me to think about the broader implications of my work. Gabriela Martinez was equally supportive, provided great feedback, and was always available for conversations. I thank both of them for the assistance throughout the process. Finally, I thank Eric Priest, who agreed to supervise an independent study on open source technology. My understanding is that Eric became one of the few law professors to serve on a dissertation committee at the University of Oregon.

I also thank the members of my cohort – Toby Hopp, Erica Ciszek, Francesco Somaini, Brant Burkey, and Fatoumata Sow – who were sources of inspiration, support, and friendship throughout our time in the program. In addition, Tewodros Workneh, Brenna Wolf-Monteiro, Pietro Monteiro, Randall Livingstone, Jacob Dittmer, Lauren Bratslavsky, Glenn Morris, Karen Estlund, Andre Sirois, Jolene Fisher, and Geoff Ostrove all made my time in Eugene enjoyable. I would also like to thank Kat at the University of Oregon for helping me with my first Linux install. Without that initial help, I would have probably never embarked on this adventure, and I am now happy to say that I've passed the gift along to others.

I also thank Christian Fuchs, Denise Rose Hansen, and Andrew Lockett at the University of Westminster for their support along the way. I feel incredibly lucky to have received a Fellowship at the Westminster Institute for Advanced Studies where I was able to expand and refine some of the ideas presented in this book. During my stay at Westminster, Ergin Bulut, Pasko Bilic, Sebastian Sevignani, Arwid Lund, Mariano Zukerfeld, Pieter Verdegem, and Maria Michalis all provided feedback and support. Also during my stay in London, a group of us attended the book launch of Massimo De Angelis's *Omnia Sunt Communia*, and the ideas outlined in that book provided inspiration for many parts of this project.

This project has also benefited from the generous support of the University of Nevada, Reno, which has been my home for the past five years. I thank all my colleagues at the Reynolds School of Journalism, and especially Dean Al Stavitsky and Associate Dean Donica Mensing for their support and willingness to accommodate research activities. In addition, this project was supported by a grant from the Vice President for Research and Innovation's office at the University of Nevada, Reno, and Ana de Bettancourt-Dias was most helpful in this regard. I also thank Rudy Leon of Evoke: Words For Hire for her assistance with indexing this book.

Last, but certainly not least, I would like to thank my family. Each of you, in your own way, provides the source of inspiration for this work. I thank my mother and father for their years of support. Without you, none of this would be possible, and I am forever indebted. My sister, Ann, inspires a wealth of adjectives: congenial, affable, good-humoured, gregarious, etc., and I'm thankful for the years of fun we've shared. My son, Caden, was only two years old when this journey began, and he is now well on his way to becoming a young man. I've enjoyed watching you grow over these years, and I look forward to many more. You are a blessing. Finally, my love and inspiration, Roberta, *meu coração*. You have truly been wonderful throughout all these years. You are a gift.

For the community, the commoners, Caden and meu coração.

CHAPTER 1

Introduction: Open Source Software and the Digital Commons

In March of 2012, The Linux Foundation released a report entitled, 'Linux Kernel Development: How Fast it is Going, Who is Doing It, What They are Doing, and Who is Sponsoring It'. The kernel is an essential part of an operating system that facilitates communication between computer hardware and software, and the Linux kernel development project is considered 'one of the largest cooperative software projects ever attempted' (The Linux Foundation, 2012: 1). Aside from a technical overview of how kernel development has changed over time, the authors included a curious note in the report's highlights: Microsoft was one of the top 20 contributors to the kernel. This marks the first time that Microsoft appeared as a top contributor, but it was not the only corporation in the top 20. Other corporate contributors included Intel, IBM, Google, Texas Instruments, Cisco, Hewlett-Packard, and Samsung, as well as others. The Linux operating system is a form of Free (Libre) and Open Source Software, or FLOSS, which allows users to freely study, use, copy, modify, adapt, or distribute the software. Why, then, would major corporations contribute directly to a FLOSS project, especially when that project seemingly does not directly contribute to corporate profits? This question becomes even more curious when one considers that many of the companies contributing to the kernel not only compete with one another in the market for information technology, but that companies like Microsoft and Google are direct competitors with Linux in the market for operating systems.

Indeed, Steve Ballmer, the Chief Operating Officer of Microsoft, once referred to Linux as 'a cancer that attaches itself in an intellectual property sense to everything it touches' (Greene, 2001). Ballmer was referencing the GNU General Public License, or GNU GPL, which is the most commonly used free software license. The GPL grants users of GPL-protected software the right to study, use, copy, modify, or adapt the software as they wish. In addition, users are granted the right to redistribute the software, as well as a modified version, and the user

How to cite this book chapter:
Birkinbine B. J. 2020. *Incorporating the Digital Commons: Corporate Involvement in Free and Open Source Software*. Pp. 1–32. London: University of Westminster Press. DOI: https://doi.org/10.16997/book39.a. License: CC-BY-NC-ND 4.0

may even charge a fee for the modified version, provided that the distributor does not place greater restrictions on the rights granted by the GPL. The GPL does not preclude corporations from modifying free software or charging a fee for their modified versions, but the corporation must still grant free software rights to end users. Ballmer's quote implies that free software is antithetical to commercial software companies. If this were the case, then Microsoft and other commercial software firms would have no incentive to contribute directly to one of the largest open source projects.

Furthermore, consider the fact that Ballmer made his denunciation of Linux on 1 June 2001. Merely 27 days later, on 28 June 2001, the United States Department of Justice found Microsoft guilty of monopolistic business practices in violation of the Sherman Antitrust Act primarily for bundling its Internet Explorer web browser with its Microsoft Windows operating system to rapidly increase its share of the market for web browsers. However, Microsoft has dramatically changed its position on Linux and open source since 2001, as signified by its inclusion in the top 20 contributors to the Linux kernel in 2012. That same year, Microsoft created Microsoft Open Technologies, Inc., a wholly owned subsidiary dedicated to facilitating interoperability between Microsoft and non-Microsoft technologies, while promoting open standards and open source. What changed during this 12-year period that Microsoft would so dramatically reposition itself in relation to FLOSS?

Microsoft is not alone. Indeed, corporate involvement in FLOSS has been increasing, especially since about 2007–2008. Table 1.1 provides an illustration of the companies that contributed to Linux kernel development for versions 4.8–4.13, which were released in 2017. The annual report for kernel development that year identified 225 companies that contributed to the project. While the Linux kernel is just one example of a FLOSS project to which corporations are contributing, other examples exist as well. This begs the question as to what motivates these companies to contribute to FLOSS projects. Furthermore, in what ways are they contributing to FLOSS projects? How do communities of FLOSS developers negotiate corporate involvement in their projects? Do communities of FLOSS developers have any recourse for unwanted corporate involvement or influence in their projects?

1.1. The Argument and Plan for the Book

The overall purpose of this book is to investigate the seemingly contradictory relationship between FLOSS communities and for-profit corporations. Working from a critical political economic perspective, I investigate the power dynamics that exist between communities of FLOSS developers and the corporations that sponsor FLOSS projects or appropriate the software production of FLOSS labourers. After all, FLOSS products and the productive process that make those products possible have been widely lauded as revolutionary changes that

Table 1.1: Top Companies Contributing to the Linux Kernel, Versions 4.8–4.13 (Corbet & Kroah-Hartman, 2017: 14).

Company	Changes	Percent
Intel	10,833	13.1%
none	6,819	8.2%
Red Hat	5,965	7.2%
Linaro	4,636	5.6%
unknown	3,408	4.1%
IBM	3,359	4.1%
consultants	2,743	3.3%
Samsung	2,633	3.2%
SUSE	2,481	3.0%
Google	2,477	3.0%
AMD	2,215	2.7%
Renesas Electronics	1,680	2.0%
Mellanox	1,649	2.0%
Oracle	1,402	1.7%
Huawei Technologies	1,275	1.5%

enable greater degrees of freedom and autonomy on behalf of users and contributors (Benkler, 2006; Raymond, 2000; Stallman, 2002). This project intervenes in these debates by tempering these claims. I position technology as a site of social struggle, and I contextualise commons-based peer production within a broader social context to illustrate how such production intersects with capitalist production. I do this by demonstrating how the purportedly revolutionary changes brought about by FLOSS and commons-based peer production are now becoming incorporated into corporate strategies and corporate structures.

The central argument presented here is that free and open source software is dialectically situated between capital and the commons. On the one hand, communities of programmers are actively working to create software as digital commons that can be accessed, used, adapted by others. By developing software iteratively this way, the pace and scale of software production increases. This represents a virtuous cycle whereby an association of software programmers actively contribute to a community that claims collective ownership over FLOSS projects. As such, FLOSS programmers can be framed as commoners insofar as they remain committed to ensuring the reproduction and sustainability of commons-based software projects over time. On the other hand, capital attempts to capture the value being produced by FLOSS communities. This includes harnessing the processes (i.e. the collective labour, or commons-based peer production power) involved in FLOSS production as well as commodifying the products (i.e. specific FLOSS projects), which can provide a basis upon which to commercially exploit the collaborative production occurring in FLOSS communities.

This is not to say that the goals of the free software commoners and capitalist firms are always antagonistic. At times they are mutually beneficial, and researchers have demonstrated how commercial sponsorship of FLOSS projects tends to make those projects more likely to attract developers and, therefore, ensures the project's longevity (Santos, Kuk, Kon and Pearson, 2013). However, we also have other examples of these relationships breaking down, particularly when it concerns the unwanted encroachment of capital upon commonly held resources like the digital commons. In these situations, the interests of the FLOSS community diverge from those of a commercial sponsor, and the relationship becomes antagonistic. The FLOSS community is faced not only with the challenge of ensuring that their digital commons remain viable, but also with ensuring that the project maintains the sense of community that enabled the project to grow in the first place. How, then, to negotiate the relationship between their digital commons and the unwanted intrusion by capital into their projects? There are a variety of factors to consider when attempting to negotiate this relationship, and the subsequent chapters provide empirical evidence for how these dynamics manifest.

The commons, generally, and the digital commons, more specifically, can be understood as an alternative system of value that is emerging from within capitalism. At times, circuits of commons value can intersect with capital accumulation circuits. Therefore, understanding the relationship between free software and capital dialectically is useful for accounting for the contradictions between these two forces that operate according to differing logics. Chapter 2 outlines these differences more specifically by drawing on theories of capitalism, digital labour, and the commons. The purpose is to develop a critical theory of the digital commons by incorporating a critique of capitalism within theories of the commons.

In Chapters 3–5, I provide three detailed case studies that illustrate different aspects of the dynamics between FLOSS communities and corporations. I separate my discussion of corporate involvement in FLOSS into three thematic areas, with each case study providing an exemplary case of these themes. The three themes are *processes, products,* and *politics*. When considered together, these three case studies are indicative of more general tendencies of corporate involvement in FLOSS projects. Furthermore, each case study offers a nuanced understanding of the complex way these dynamics work, and they allow for a detailed unpacking of some of the contradictions inherent in the relationships.

To begin, Chapter 3 focuses on Microsoft's contentious relationship with FLOSS. This relationship is indicative of the ways in which the *processes* involved in FLOSS production effectively ushered in a new era of industrial software production. While other companies demonstrated a willingness to cooperate with FLOSS communities, Microsoft's dominance of the software market for personal computing during the 1980s and 1990s makes it an instructive case for understanding how software production changed over time. The major historical event here is the antitrust ruling against Microsoft, which marked the end of an era in which software production was largely accomplished within a single firm that sought to exclude others from accessing its code. Indeed, one of the consent decrees in the Microsoft antitrust ruling was that Microsoft provide third parties access to its application programming interfaces (APIs). This was a radical departure from Microsoft's earlier practices, whereby the firm rose to power by using anticompetitive business practices.

Coinciding with Microsoft's dominance of the software market and its eventual antitrust conviction in the 1990s were other software firms trying to find a way to transform FLOSS products into successful commercial products. My analysis of Red Hat, Inc. in Chapter 4 is indicative of how FLOSS *products* get incorporated into a commercial firm's overall business strategy. Red Hat remains the largest and only publicly traded company providing software and services that are completely based on free software. As such, Red Hat cannot rely on traditional copyright protections to exclude others from using the underlying source code included in its software. Thus, my analysis of the firm explores how Red Hat has been able to create a profitable business based on free software.

Finally, the third case study in Chapter 5 focuses on how FLOSS communities cope with unwanted corporate influence in their projects. Sun Microsystems was an important corporate sponsor of FLOSS projects, but it was acquired by the Oracle Corporation, which had different plans for those projects. In that chapter, I focus on the diverse destinies of three such projects – the OpenSolaris operating system, the MySQL relational database management system, and the OpenOffice productivity software – and the ways that the communities involved in those projects resisted Oracle's encroachment into their projects. In effect, the case study illustrates the *politics* involved in negotiating boundaries between FLOSS communities and corporations, while also demonstrating some of the strategies FLOSS communities can use to protect their projects.

In the remainder of this introduction, I provide more context for understanding the significance of FLOSS. This includes historically situating FLOSS within a broader discussion of the commons, as well as some of the key historical moments in the development of software, generally, and FLOSS, more specifically. In each of these sections, I also offer some notes on the terminology used throughout the book, which will hopefully assist in avoiding conceptual confusion. Following those sections, I discuss the cultural significance of FLOSS. I conclude the chapter with a note on the methodology used for the current study. Readers who are already familiar with the history of FLOSS and its defining characteristics may wish to skip directly to the next chapter or the note on methodology at the end of this chapter.

1.2. Situating Free (Libre) and Open Source Software

Although free software and open source communities are related and, in some cases, not mutually exclusive, each of them has distinct characteristics that can best be described by reference to the ethos underlying each movement. To contextualise the emergence of FLOSS within the evolution of the computing and software industries, a brief history of these industries is provided below. Following that discussion, I focus on situating two key figures associated with FLOSS within their historical context: Richard Stallman and Linus Torvalds. These two figures represent free software and open source, respectively.

1.2.1. Historicising Free and Open Source Software

The use of machines for processing information or calculating differences in numbers, human beings performed such work. But human calculations were, at times, prone to errors. To reduce this uncertainty, Charles Babbage, a philosopher and mathematician working at the University of Cambridge in 1822, proposed that it was 'only by the mechanical fabrication of tables that such errors can be rendered impossible' (Gleick, 2011: 95). Such was the proposition for Babbage's Difference Engine, which performed routinised calculations mechanically, and was arguably the genesis for modern computers as we know them today. Later, Babbage expanded on his idea and planned a new type of machine that was capable of being controlled by instructions that could be encoded and stored to facilitate operation. The new iteration of the idea was called the Analytical Engine, but this still only provided the idea for the hardware or mechanisms necessary for such processes to occur. What was needed for this hardware was software.

The idea for software arguably originates with Augusta Ada Byron King, the Countess of Lovelace, otherwise known simply as Ada Lovelace. In 1843, she developed the idea that Babbage's Analytical Engine could perform a series of operations beyond the mere calculation of numbers. By abstracting from the differences between two things, Lovelace posited that the Analytical Engine could be programmed to perform operations that relied on symbols and meanings, which, in turn, could be communicated to the machine. Although Lovelace's idea was never realised in her lifetime, she is credited with developing the idea for software and is known as the first programmer.

While Babbage and Lovelace are credited as pioneers in developing the ideas for modern computers and software, the construction of such machines did not begin until World War II. Developments in the field of computer science and information theory – like Kurt Gödel's incompleteness theorem, Alan Turing's idea for a Universal Turing Machine, Claude Shannon's mathematical theory of communication, and Norbert Wiener's cybernetics – provided the intellectual inspiration for the development of such machines. Before, during, and after

World War II, many of the developments leading to modern computers were used for military purposes. Most notable, perhaps, were the German Enigma machine that was used to encrypt secret messages and the electromechanical bombes used by the United Kingdom to decipher those messages (Smith, 2011). However, in 1941, Konrad Zuse, a German electrical engineer, built the Z3, which is regarded as the first electro-mechanical, programmable, fully automatic digital computer (Zuse, 1993). The first comparable computer in the U.S. was developed by John Atanasoff at Iowa State University in 1942 (Copeland, 2006). Only one year later, the first fully functioning electronic digital computer was put to use by the cryptanalysts working at Bletchley Park in the U.K. as part of the Government Code and Cypher School. The Colossus, as the new machine was known, was programmed to decipher German communications during the war. By the end of the war, Bletchley Park had 10 Colossi working to decode German communications (Copeland, 2006).

Following these initial landmarks, the development of modern computers accelerated as many of the early pioneers began working for academic institutions and private companies after the war. In the United States, Grace Hopper, who served in the United States Navy Reserves as a member of the Women Accepted for Voluntary Emergency Service (WAVES) during World War II, was assigned to the Bureau of Ships Computation Project at Harvard University. While there, she worked on the Mark I computer project, which was built by IBM in 1944. Later, after she began working for private companies, Hopper popularised the idea of machine-independent programming languages. This led to the development of the Common Business-Oriented Language (COBOL) in 1959. Hopper is also credited with popularising 'debugging' as a term for removing defective material or code from a program. While Hopper may not have invented the term, she popularised it by literally removing a moth from a Mark II computer at Harvard University after it had caused the machine to short circuit (Deleris, 2006).[1]

During the 1960s, the creation of microprocessors drastically reduced the cost of computing. As a result, communities of hobbyist programmers and computer enthusiasts began to experiment with the technology in the following years. One notable example was the Homebrew Computer Club, started by Gordon French and Fred Moore in 1975 at the Community Computer Center in Menlo Park, California. The club provided an open forum for hobbyists to trade parts and advice about the construction of personal computers. The goal was to make computers more accessible to others. More will be said about this specific hobbyist community in Chapter 3, as it played an important role in the rise of Microsoft. Aside from these hobbyist communities, the majority of computer development occurred within the military, academic institutions, and private companies.

Most notable were the initial developments within the Defense Advanced Research Projects (DARPA), which was created in 1958, as well as the Artificial Intelligence Lab at the Massachusetts Institute of Technology (MIT), which was founded in 1970.[2] Programmers working at the time were using a proprietary programming language called Unix, the intellectual property rights for which were owned by AT&T. One of the programmers working at MIT was Richard Stallman, who began working in the lab in 1971. Stallman found that when he wanted to work with the Unix programming language outside of officially sanctioned spheres, he was denied access to the code by AT&T. In protest, he posted messages to computer-based bulletin boards in 1983 announcing that he was developing a Unix-based language that would be available for free so that others could use the language however they saw fit. In 1985, Stallman published 'The GNU Manifesto', which outlined the goals of his new project, his reasons for developing the project, and what the project was aimed at fighting back against.[3] The programming language was called 'GNU', a recursive acronym standing for 'Gnu's Not Unix'. Along with the programming language, Stallman developed the GNU Public License (GPL), which stipulated that anyone could access the source code for free, and that anyone using the GPL agreed to make their contributions available under the same conditions. This would ensure that computer programmers could freely share their work with one another, thereby creating a common form of property that developed in opposition to its proprietary and closed counterparts.

Stallman became the figurehead of the movement against proprietary software. He viewed access to source code as a fundamental right, which he wanted others to believe in as well. He summed up this view in his famous dictum, 'Free as in freedom, not as in free beer', thus positioning free software as a moral right (Stallman, 2002). The free software definition stipulates that 'users have the freedom to run, copy, distribute, study, change and improve the software' (Free Software Foundation, 2012). As the principles of free software grew beyond the borders of the U.S., others have tried to reduce the confusion over the English term 'free' by using the French term *libre* rather than *gratis*. Stallman established the Free Software Foundation (FSF) to promote his movement against proprietary software, and he represents an impassioned counter-cultural figure who continues to espouse his free software philosophy.

While Stallman is generally considered to be the figurehead of the free software movement, open source software is generally associated with Linus Torvalds. In many ways, Torvalds and Stallman have similar stories, but differ on philosophical terms. During the 1980s, free software projects were being developed but generally on a smaller scale. Free software had not yet found a way to coordinate efforts on a larger scale. Torvalds wanted to work on kernel development for an open-source operating system. Rather than relying on numerous programmers all working independently on such a task, Torvalds released the source code for his project, which he was calling 'Linux', a portmanteau of his name, Linus, and the language he was working with, Minix (itself a simplified derivative of AT&T's Unix). Torvalds suggested that anyone who was interested in contributing to such a project was encouraged to do so, if they released their work back to the community so that others could progressively work toward completing the kernel. The project proved to be successful, and eventually led to the creation of the open source operating system, Linux. Coordinating such a large-scale programming project was accomplished by asking those working on the code to release their work, no matter how small the changes seemed. The rationale was that coordinated efforts reduce the amount of redundant work, which was summed up in the adage 'with many eyes, all bugs are shallow', which Eric Raymond refers to as 'Linus's Law' (Raymond, 2000).

Stallman and Torvalds differ with respect to how they view the relationship between free software and proprietary software. Whereas Stallman tends to be somewhat more confrontational in his opposition to proprietary software, Torvalds is less so. Williams (2002) describes a decisive moment at a conference in 1996 where Stallman and Torvalds appeared on a discussion panel together. Torvalds expressed admiration for the work that Microsoft was doing and suggested that free software advocates could work together with companies. Such a suggestion was generally seen as taboo since Stallman was perceived with esteem by the programming community, and the Free Software Foundation generally took a very adamant stance against proprietary software companies. Powell (2012) frames this distinction between free software and open source similarly:

> open source software as an industrial process grew out of the culture of free software development, but departed from the latter's political focus on the value of sharing and the maintenance of a knowledge commons, and instead focused on the efficiency of open source processes for software production (692).

This moment at the 1996 conference thus marked a watershed moment in which the fervour of the free software movement thawed a bit, as Torvalds came to represent a more liberal approach to free software. By 'liberal' here, I am referring to the literal definition rather than a specific political position; the term should be understood as something that indicates an openness to new perspectives or behaviours while willing to abandon traditional values. In this regard, Linus's expression of support for the work that Microsoft was doing signalled an openness to working with Microsoft (or other commercial firms) simply to produce the best software rather than an adherence to the anti-corporate stance of Stallman and the Free Software Foundation.

In sum, then, we can understand the free software and open source movements with respect to these differing philosophical positions. Stallman and free software advocates tend to make moral claims against supporting proprietary software, while Torvalds and open source tend to be associated with a more liberal and inclusive stance. While Stallman and Torvalds have been used to illustrate the differences between free software communities and open source communities, they should not be viewed as mutually exclusive communities, nor should they be seen as representative of the entire free software and open source communities. One of the peculiarities of the free and open source software community is that, although the overall community is united in their belief that software ought to be free for users to study, modify, adapt, or customise, its members will often vehemently defend their preferred free software project while deriding others. In a sense, this signals to others where their loyalties lie and engenders stronger ties within niche communities that exist within the larger FLOSS community. The present project is less concerned with these intra-group fissures than the relationship of the community to the corporations that rely on their labour. To that end, the combined term 'Free (Libre) and Open Source Software' or 'FLOSS' is used to refer to the overall community.[4]

1.2.2. The Unseen Ubiquity of Free and Open Source Software

From its beginnings in the 1980s and 1990s, FLOSS has proved to be an efficient and effective way of producing software. Whether we realise it or not, most of us rely on FLOSS in our everyday computing, as it provides critical infrastructure that enables the Internet to function. As an example of the size and scope of some FLOSS projects, consider the Linux kernel, which was discussed in the introduction to this chapter. When it was first released in 1991, the Linux kernel featured approximately 10,000 lines of code. Version 4.13 of the Linux kernel was released in September 2017 and featured nearly 25 million lines of code, which was produced by nearly 1,700 developers and 225 companies (Corbet and Kroah-Hartman, 2017: 11). Furthermore, Linux has become widely used as an operating system. For example, Linux (or other operating systems derived from Linux) holds 100% market share in the market for supercomputer operating systems (Top500.org, 2018a). These computers are the most powerful computers in the world, and all of them rely on Linux or Linux-based operating systems. This includes the United States Department of Energy's supercomputer at the Oak Ridge National Laboratory in Oak Ridge, Tennessee, which at the time of writing is home to the world's fastest and most powerful supercomputer (Top500.org, 2018b).[5] While Linux does not yet have a significant share of the personal computing desktop market, the operating system has been customised and used within a variety of contexts.

Within the United States, Linux is used for high-level military operations. For example, the United States Navy announced that its $3.5 billion warship, the USS Zumwalt, which has been described as 'the most technologically advanced surface ship in the world', will effectively serve as an armed floating data centre that features server hardware running various Linux distributions and more than 6 million lines of code (Mizokami, 2017, Gallagher, 2013). In addition, the International Space Station switched from the Windows operating system to Debian Linux, according to Keith Chuvala, the Manager of Space Operations Computing at NASA, because they wanted to have '...an operating

system that was stable and reliable – one that would give us in-house control'
(Bridgewater, 2013).

Indeed, Linux and Linux-based systems also provide essential components
for some of the most recognisable technology companies, which was discussed
briefly at the beginning of this chapter. Despite the fact that I have only selected
a few companies for detailed examination in the subsequent chapters, one
could find other similarly intriguing case studies that would exemplify different
dynamics between corporations and FLOSS communities. As such, it is worth
mentioning some notable examples here simply to emphasise the ubiquity of
Linux. Google's Android operating system, for example, is one of the world's
most popular mobile platforms, and it is based on the Linux kernel. However,
there are certain key components of the Android operating system that remain
proprietary to Google (see Amadeo, 2018). Aside from Google, other companies like Canonical rely on Linux for creating customised operating system distributions. Canonical produces Ubuntu, which is one of the most widely used
Linux distributions.

Linux has also seen widespread adoption around the world. Some countries
have developed their own versions of Linux to meet specific needs, and some
cities have even required that Linux be given preference over other operating systems. For example, between 1999–2001, four cities and municipalities
in Brazil – Amparo, Solonópole, Recife, and Ribeirão Pires – passed laws that
required government agencies to use or give preference to Linux (Tramontano and Trevisan, 2003; Festa, 2001). The decision to switch to free software
systems was mainly economic, as Brazil reported spending nearly $1 billion
on software licensing fees to Microsoft between 1999–2004 (Kaste, 2004). By
switching to free and open source software, Brazil estimated that they could
save approximately $120 million per year (Kingstone, 2005). Brazil remains one
of the more progressive countries in its support of free software (see Birkinbine,
2016a; Schoonmaker, 2018; 2009). Many of the country's policy measures and
initiatives related to FLOSS have been driven by communities of activists who
have been able to intervene in policymaking processes to institute policies that
seem to contradict the prevailing neoliberal ideology. In an excellent article on
the subject, Shaw (2011) framed these activists as *insurgent experts*.

Similar measures to support free software were taken in Kerala, India, as the
state adopted a policy to remove proprietary software from its educational system. According to one estimate, the switch saved the state of Kerala roughly
$58 million each year (Prakash, 2017). The German city of Munich developed
its own version of Linux called LiMux (Linux in Munich), which it used as an
operating system for its 15,000 city council members before announcing a shift
back to Microsoft in 2017 (Heath, 2017). The National University of Defense
Technology in China has also developed its own Linux-based operating system
called Kylin. In addition, the computers used for the One Laptop Per Child
project, which was founded with the goal of bringing low-cost computers to
developing countries for educational purposes, featured a free and open source

operating system based on Fedora, the free software project sponsored by Red Hat, Inc., which will be discussed in Chapter 4.

Beyond the increasing use of Linux, open-source principles have been used in areas outside of information technology. For example, open source hardware (see Söderberg, 2011) can increase access to physical goods, including furniture, musical instruments, construction materials, and wind turbines for generating renewable energy. Such projects are particularly attractive to those living in developing countries, where access to information, goods, and services may be restricted or limited. One of the more ambitious projects in this area is the Open Source Ecology project, which offers 'open source blueprints for civilization,' and includes instructions for building industrial machines with recycled or low-cost materials (Open Source Ecology, 2019). While this is just one notable example, it demonstrates the optimism and creativity involved in applying open source principles to a whole way of living rather than simply information technology. However, the core values inherent in these projects do not necessarily originate in open source software. Rather, the cultural values of openness, sharing, mutual aid, respect, and conviviality are foundational values for building a community. When applied on a broader scale, these principles hold the promise of a more sustainable future, especially when such principles are linked with environmental and ecological preservation practices. But these principles only become radical propositions in a system that discourages or provides little incentive for valuing them.

Despite the fact that FLOSS communities comprise a socio-technical system insofar as their activities are made possible by and exist within a technologically mediated realm, FLOSS enthusiasts also congregate and cooperate in-person through a network of Linux User Groups (LUGs) around the world. Regular meetings of LUGs are held to promote FLOSS, to assist new users with installing FLOSS, to troubleshoot any issues that may arise when using FLOSS, or to simply meet other people interested in FLOSS. In this sense, the social connections that exist within these groups are mediated by their mutual interest in technology. Because members of the FLOSS community are brought together by their mutual appreciation of technology, their cultural practices depend upon and are supported by interconnected network technologies. As more people become connected to the network, the opportunities for additional participants in these communities grow.

One final point deserves attention here too. It seems like an increasing amount of our social lives is spent on the Internet where we work, communicate with friends and colleagues, read news, watch movies and television, and listen to music, among other activities. When we connect to the Internet and visit websites, our requests for information are relayed through a network of interconnected servers that facilitate communication between other clients on the network. The operating systems running those servers are increasingly FLOSS projects like Linux or FreeBSD, but Microsoft also designs server software. This provides another example of FLOSS projects competing with proprietary companies like Microsoft. Consequently, and whether we realise it or not, our ability to connect to the Internet may depend, in part, on the ability of FLOSS projects to work together with proprietary software. This further demonstrates the need for understanding the ways in which proprietary software and FLOSS projects work together, as well as what happens when these relationships break down. Unpacking the dynamics that exist in these relationships can help us understand either the enabling or constraining of our ability to connect with others online.

What these examples should illustrate is that Linux but also FLOSS more generally has become more than just a tool used within the computer hobbyist community. Its widespread and increasing adoption across the globe within a variety of high-level contexts demonstrates the power of the FLOSS production model as well as the effectiveness of its products. As FLOSS continues to be used within an increasing variety of contexts, understanding the ways in which corporations, governments, non-profit organisations, and other types of institutions are involved in FLOSS projects will become increasingly important. Therefore, FLOSS provides an important area for research not just because of its increasing ubiquity, but also because of the claims that have been made about the democratic, egalitarian, and non-market characteristics of its products and processes. This is precisely how this project seeks to contribute to such debates.

1.2.3. FLOSS and Hacker Culture

The term 'hacker' has taken on negative connotations recently, but the term is generally used to describe anyone who 'tinkers' with or makes changes to technology to create something new. Steven Levy (1984) outlined the principles of the hacker ethic. Among other elements, Levy claimed that computers can be used for creative purposes, hackers ought to be judged by the quality of their work rather than any other characteristic (gender, race, ethnicity, etc.), and that having the ability to hack is a prerequisite for hacking. This last caveat may seem obvious but, in order to perform a hack, a hacker must have access to the technology (in this case, the source code). In other words, closed, proprietary technologies that do not allow for tinkering may be viewed as unjust.

Indeed, when faced with closed, proprietary, or otherwise secured technologies, a hacker may attempt to circumvent or remove those restrictions. At times, this is done to make a point about information security, but it is also done to signal to others that they deserve credit for the sophistication of their hack. This signalling motivation is also recognised within open source software communities (Lakhani and Wolf, 2005), especially because FLOSS programmers are interested in remixing, modifying, adapting, or creating something new from a given product. The same signalling motivation has been used to understand why programmers contribute to FLOSS projects. Lakhani and Wolf (2005) explain that signalling can take place within at least a couple of levels. At the level of the individual, a single hacker may perform a hack to signal his or her skills to others. Hackers might also use this type of signalling to communicate their skills to potential employers to secure paid employment. Gaining recognition within the broader community for performing certain programming tasks effectively can translate into increased job opportunities with companies looking for specific skills.

However, a different type of signalling takes place between groups of hackers. Groups or collectives may signal their prowess to others by shutting down a web site or otherwise disrupting services. Often, this is done in the spirit of competition, but can also be explicitly driven by a particular ideology. For example, nationally based hacker groups can be found in Syria where a pro-Syrian government hacking group called the Syrian Electronic Army has waged hacking battles against the pro-rebel hackers associated with the Free Syrian Army (Fitzpatrick 2012). In these situations, hacker groups strategically target the web sites of their opponents to signal the strength of their movement.

Although the signalling appears to be the most prevalent motivation, Weber (2004) identifies other motivations as well. In a survey of self-identified hackers, respondents reported their primary motivation for contributing to FLOSS development was a desire to challenge oneself and perform creative work. This seems to support what Levy (1984) identified as primary tenets of the hacker ethic: creativity and aesthetics. Weber (2004) also found additional motivations reported in the survey, including the belief that all software should be free, which echoes the philosophy of Richard Stallman and the Free Software Foundation. Weber concludes that motivations are diverse and that the results from these surveys need to be properly contextualised. For instance, many contributors to FLOSS development do not disclose their identity or any institutional affiliation. Indeed, a look at the credits file for users contributing to the development of the Linux kernel shows that most contributors are listed in the 'unknown' category. This means that a large portion of the FLOSS community simply chooses not to self-identify. Therefore, the results of any survey that claims to represent the entire FLOSS community must be approached somewhat sceptically.

While signalling and creativity are certainly important factors for understanding the motivations of hackers and FLOSS contributors, my own view is that the most robust scholarship on the cultural significance of free software and FLOSS production comes from Christopher Kelty. Kelty (2008) positions free software as a *recursive public*, which he defines as:

> a public that is vitally concerned with the material and practical maintenance and modification of the technical, legal, practical, and conceptual means of its own existence as a public; it is a collective independent of other forms of constituted power and is capable of speaking to existing forms of power through the production of actually existing alternatives (Kelty, 2008: 3).

In other words, in the process of actively contributing to FLOSS projects, FLOSS programmers actively create, recreate, or reproduce the infrastructure that enables their activity to take place. This has conceptual links with other theories of the commons that position the commons as a process or a way of becoming (Dyer-Witheford, 2006; Linebaugh, 2008; Singh, 2017). Similarly, Rossiter and Zehle (2013) argue the commons are not purely 'given as a fragile heritage to be protected' against enclosure, but they must be actively constructed. FLOSS communities actively produce the digital commons as code, which is produced and licensed under intellectual property licenses that permit users to use the code and adapt it for their own purposes. These alternative intellectual property licenses take many different forms. The original copyleft licence to see widespread use was the GNU General Public License.[6] Other notable examples are the Creative Commons[7] licences, which allow varying levels of use for the protected property under conditions set by the creator. For example, users may make their creation freely available and permit others to use it, if those users provide attribution to the original author.

Kelty (2008) furthermore claims that FLOSS programmers 'do not start with ideologies, but instead come to them through their involvement in the practices of creating Free Software and its derivatives' (7–8). Coleman (2004) makes similar claims when she refers to the 'political agnosticism' of FLOSS. The complex forces at play in this agnosticism stem from an outward denial of specific political affiliations even while 'political denial is culturally orchestrated through a rearticulation of free speech principles, a cultural positioning that simultaneously is informed by the computing techniques and outwardly expresses and thus constitutes hacker values' (Coleman, 2004: 509). Coleman continues by explaining that the core of the moral philosophy espoused by the FLOSS community is a 'commitment to prevent limiting the freedom of others' (509). This utilitarian ethic of openness is what is necessary for FLOSS programmers to continue building state-of-the-art computer programs because it is precisely the ability to tinker, adapt, and improve upon software that enables innovation to occur within software development.

These principles, as well as the outward denial of a specific political position, are, in part, what has enabled the FLOSS community to attract such a large community. Of course, this is not to say that all members of the FLOSS community reject specifically political ideologies. One needs to look no further than Eben Moglen's (2003) 'dotCommunist Manifesto', which offers a polemic against the regimes of private property. Indeed, he concludes the manifesto with the following seven principles in the struggle for 'free speech, free knowledge, and free technology' as well as a concluding note on how this struggle will bring about a more just society:

1. Abolition of all forms of private property in ideas.
2. Withdrawal of all exclusive licences, privileges and rights to use electromagnetic spectrum. Nullification of all conveyances of permanent title to electromagnetic frequencies.

3. Development of electromagnetic spectrum infrastructure that implements every person's equal right to communicate.
4. Common social development of computer programs and all other forms of software, including genetic information, as public goods.
5. Full respect for freedom of speech, including all forms of technical speech.
6. Protection for the integrity of creative works.
7. Free and equal access to all publicly produced information and all educational material used in all branches of the public education system.

By these and other means, we commit ourselves to the revolution that liberates the human mind. In overthrowing the system of private property in ideas, we bring into existence a truly just society, in which the free development of each is the condition for the free development of all.

(Moglen, 2003)

Similarly, Dmitry Kleiner's (2010) *Telekomunist Manifesto* outlines proposals for developing a working class politics online. His proposals for venture communism as well as a copyfarleft licensing regime offer concrete proposals for developing alternatives within existing frameworks, but doing so in a way that is guided by radical politics. Both of his proposals are aimed at preserving and protecting the commonly held property of independent producers from capitalist exploitation or co-optation.

It is precisely because the collective productive activity of the FLOSS community is so valuable for software production that capitalist firms are interested in harnessing this power. At the same time, this is also the reason that critical scholars like Kleiner have sought ways to preserve that value within the communities who create such value, even if they offer different proposals for how to do so. Taken as a whole, then, this community holds tremendous value for software production. The authors discussed above, particularly the work of Kelty (2008) and Coleman (2004; 2013), offer some of the best work for understanding the cultural significance of FLOSS as well as the ethics underlying the FLOSS community. However, there is still the pressing question of what happens when the specific cultural, political, and economic values of the FLOSS community intersect with circuits of capital accumulation. This was one of the tensions that Kleiner (2010) was trying to address when developing his proposals for alternatives. Moreover, in what ways does the FLOSS community negotiate and justify the dual position of advocating for open knowledge and market success simultaneously? Some of the best work exploring the complex set of dynamics at work in this regard has been that of Alison Powell (2012; 2016; 2018). In exploring the ways that participants in peer production communities negotiate competing moral visions for their projects, Powell (2018) argues that participants often engage in 'operational pragmatics' that are used to justify various design decisions. In doing so, participants collapse distinctions between advocacy for open knowledge and market success even if these distinctions seem to be at odds with one another. In effect, both are viewed as 'good' or virtuous, that function as 'regimes of justification' when making decisions about design (Powell, 2018: 514).

How, then, can we understand these complex and intertwined ways of negotiating cultural differences both within peer production communities as well as their intersection with capital accumulation circuits? Is it possible for peer production communities to be exploited by capital if they are willing participants in designing products for market success? After all, corporations are keenly interested in harnessing the productive power of the FLOSS community. The following section discusses one way to theorise the ways in which companies relate to FLOSS communities. However, the following chapter will discuss these specific dynamics in greater detail by drawing from theories of capitalism, digital labour, and the commons, while exploring the ways in which exploitation occurs when capital and the commons intersect.

1.3. Open Source Business Models

The previous section demonstrates how the specific cultural dynamics at play in FLOSS communities have been explored quite effectively by other scholars, including the significance of those dynamics for cultural production more broadly. However, the economic arrangements between corporate firms and FLOSS communities have been explored comparatively less. This book aims to offer some greater descriptive detail as to how these dynamics specifically manifest as FLOSS communities and corporations negotiate the boundaries between their respective organisations. However, one attempt to develop a typology of open source business models is worth mentioning here.

As part of their broader treatment of open source software, Deek and McHugh (2008) develop a typology of open source business models. The typology contains five different models that have been used in trying to profit from FLOSS. Table 1.2 provides an illustration of this typology, providing the types of business strategies employed, a description of the strategy, and an example of a company or product that is representative of the strategy.

The first business model relies on dual licensing, in which the owner of copyrighted software provides free and open distributions for non-profit users but requires for-profit customers to pay a fee to use the software. The exemplary case here is MySQL, which is an open source database management system. The company provides a free version of its software under the General Public License (GPL), which stipulates that any derivative software using the GPL-licensed software must also be made available under the same licence. MySQL also provides an advanced commercial version of its software to for-profit corporations, which can be customised to the users' specific needs or integrated with that company's proprietary software.

Table 1.2: Types of Open Source Business Strategies, adapted from Deek and McHugh (2008: 272).

Business Strategy	Description	Examples
Dual Licensing	Owner of copyrighted software provides a free and open distribution for non-profit users but requires for-profit customers to pay a fee to use the software.	MySQL
Consulting	Company assists other companies with planning, strategy, and implementing appropriate open source solutions within their business.	Olliance Consulting (division of Black Duck Software), LQ Consulting
Distribution & Services	Company provides services for non-expert computer users by handling the compilation of stable, updated, and prepackaged software suites that are distributed to users (clients).	Red Hat, Canonical
Hybrid open/ proprietary – Vertical Development	Using open source as a base upon which proprietary software can be built.	Google
Hybrid open/ proprietary – Horizontal Arrangements	For-profit company becomes directly involved in supporting open source projects to supplement its own business operations.	IBM, Microsoft

The second type of business model is one in which a company provides consulting services for FLOSS. Quite simply, companies that adopt this model assist other companies with planning, strategy, and implementing appropriate open source solutions within their business models. Among other things, Black Duck Software provides consulting services through its Olliance Consulting division.

The third business model is one in which a company provides FLOSS distributions and services, and the exemplary company here is Red Hat. Unlike MySQL, which owns the copyrights for its software, Red Hat creates and provides its own distribution of Linux. In addition, Red Hat provides training, education, documentation, and support for its Linux distribution. In other words, Red Hat provides a service for non-expert computer users by handling the compilation of stable, updated, and prepackaged software suites to be distributed to users. In some ways, then, Red Hat behaves similarly to a proprietary software provider, except that it does not own the intellectual property rights for the software it sells and services. Rather, the company sells and provides its own Linux distribution, which it can do because of the open licensing model of Linux.

Whereas the first three business models are solely related to FLOSS, the remaining two rely on a hybrid of both open and proprietary software. The fourth model is a hybrid of both proprietary and open software that relies on vertical development with FLOSS. Vertical development means using open source software as a base upon which proprietary software can be built. One of the major corporations that uses this model is Google. In fact, Google does not sell its software at all; it develops and maintains its own software in-house, while selling services provided by its software to other customers. Of course, Google's search engine is proprietary, but Google uses the Linux core to support its proprietary search services.

The final model is a hybrid of proprietary and open software, but one in which the company relies on horizontal arrangements. This is the business model that lies at the heart of this book project. In these relationships, for-profit corporations become involved in open source projects. Drawing from Fogel (2005), Deek and McHugh (2008) claim that the reasons for corporate involvement are diverse, but include everything from spreading 'the burden, cost, and risk of software development across multiple enterprises to allowing companies to support open source projects that play a supportive or complementary role to their own commercial products' (277). IBM is one example of this type of business model. For example, IBM's WebSphere application, which enables end-users to create their own applications, was built using the Apache web server, which is open source. Thus, by supporting open source projects like Apache, IBM is indirectly supporting its own interests. Furthermore, IBM directly competes with Microsoft as a platform for applications. Because IBM supports Linux, it is not only investing in the reliability of its own products but may simultaneously weaken Microsoft's market position, especially because Linux is also a direct competitor of Microsoft.

In sum, then, this section has discussed how FLOSS has been used in differing ways by drawing on the typology developed by Deek and McHugh (2008). The most fruitful area of study for the purposes of this project was the hybrid open/proprietary model that relies on horizontal arrangements, although other projects are discussed, like MySQL, which represents other types of business strategies. The corporations that rely on horizontal arrangements are most interesting because of their direct involvement in FLOSS projects. Thus, these companies need to maintain a good relationship with the broader FLOSS community. When the norms of the community are violated by a company, the community can abandon a project, which can effectively end commons-based production on the project. In this sense, the FLOSS community leverages its collective labour power against undue corporate influence in its commons-based resources. This was the case when the Oracle Corporation acquired Sun

Microsystems. This case will be discussed in greater detail in Chapter 5. For now, however, it is important to note the two different examples of companies using hybrid horizontal agreements to two different ends. In the case of IBM, the company maintained a relatively stable relationship with the open source community. In the other, Oracle overstepped its bounds by violating the norms of the community. As more and more corporations become involved in FLOSS projects, the relationships that exist between the community and the corporations that rely on their collective labour power will be subject to changes.

1.4. FLOSS as Digital Commons

The seemingly contradictory relationship between FLOSS communities and corporations is further exacerbated by the fact that FLOSS has consistently been held up as the primary example of a digital commons. In medieval England, the commons referred to a portion of land owned by the lord of the manor, which certain tenants had the right to use for their needs. These rights included the right to cultivate soil, produce crops, feed livestock, and other activities. The concept has since been expanded from this very specific meaning to encompass any resource that is owned by a community or a resource that may be accessed by a broader community of people.

In tracing the roots of scholarship on the commons, most scholars bookmark the work of Elinor Ostrom (2005; 1990). The narrative often begins with Ostrom's work, and focuses on how her ideas developed and influenced subsequent generations of scholars.

While Ostrom is a towering figure in scholarship on the commons, this simple narrative tends to obfuscate the broader history and context within which Ostrom's work is situated. Locher (2016) clarifies this history by demonstrating how Ostrom's work can be contextualised within a broader history of scholarly debates within economic, political, and anthropological scholarship concerned with the best way to achieve development. These debates were concerned with the role of the state, the market, and local communities in the project of development during the post-World War II period. This scholarship can be linked with the United States' international development projects through its flagship institution, USAID, in the 1970s–80s.

Two assumptions in the approach to development dominated this period. One was the assumption of the 'tragedy of the commons' or the fallacy of collective action, based primarily on the work of Garrett Hardin (1968). Hardin argued that the commons were ultimately unsustainable because they were at risk of overexploitation as members of the community acted in their self-interest to maximise personal gain. Thus, there was a fallacy in the logic of collective action; it was simply impossible for communities to govern collective resources without overexploiting them. The second assumption was that the liberal technocratic state ought to be the central agent in development through economic planning and coordinating large-scale development projects. This assumption was driven by the success of the New Deal and the welfare state in the post-war period. As such, the model was viewed as the primary means for developing countries in the Global South, where traditional practices would give way to modernisation to boost economic productivity.

During the 1970s, these assumptions were challenged by development anthropology, which analysed 'adaptive socio-ecological strategies' used by local communities to ensure the survival of ecological resources (Locher 2016, 313). Often, these decision-making strategies were situated within complex systems of customs and social rules that developed from local communities' historical experiences with their broader environment. Challenges to these assumptions continued in the 1980s as neoliberal economics emerged as an alternative to welfare state capitalism. Informed by rational choice theory, which privileged calculating and efficient economic decision-making by profit-maximising individuals, the goal was to unleash productive capacity in the private sector through deregulation and privatisation. Neoliberal doctrine thus argued for dismantling state regulation and withdrawing the state from social provision. As such, neoliberalism represented not just an economic doctrine but also 'an ethic in itself, capable of acting as a guide for all human action, and substituting for all previously existing ethical beliefs' (Treanor, 2005: n.p.).

It was within this context that Ostrom's scholarship, in collaboration with others, sought to illuminate the ways that local communities govern common-pool resources outside of the binary of either state provision or market relations. For example, Hess and Ostrom (2007) argued against the tragedy of the commons thesis by focusing primarily on two points: first, Hardin assumes that the sheep herders are acting according to the principles of neoclassical economics and are individually acting in their self-interest rather than allowing for forms of common governance, whereby concessions are made to the other sheep herders. Second, Hardin frames the issue within the binary choice between socialism and capitalism. However, the framing is fallacious for a couple of reasons. The commons under feudalism were owned by a private individual and not the state. Furthermore, Ostrom (1990) demonstrates how different types of commons can be governed collectively so individual short-term gains can be compromised for the long-term survival of the common resource. In effect, Ostrom (1990) provided some nuance to the way that we understand commons, especially because they were often placed in a binary opposition that was representative of Cold War-era assumptions about social development: either state provision of common property (socialism) or private property ownership (capitalism).

Ostrom focused on the diverse ways that different commons are managed by those communities that claim some sort of association to the resource. The types of common-pool resources governed in this way vary, but the initial focus was on natural resources like fisheries, grazing pastures, groundwater basins, and irrigation systems. Later, Hess and Ostrom (2007) would expand the study of the commons to non-tangible resources like knowledge and information. Table 1.3 illustrates different types of property by providing a simple matrix of two factors: rivalry and excludability. Rivalry refers to the extent to which a resource is finite or requires reproduction. Highly rivalrous goods tend to be finite objects like apples, which need to be planted again to reproduce the crop, while low rivalry goods tend to be intangible goods that can be reproduced without much additional cost, like ideas, information, or knowledge. Excludability refers to the extent to which an owner of such goods can exclude others from accessing or using that good. Highly excludable goods are protected by private property rights, whereas goods with low excludability may be used by anyone. Following from these terms, the matrix for rivalry and excludability would look something like this:

Table 1.3: Typology of Property, adapted from Hess and Ostrom (2007) and Frischmann (2012).

		Excludability	
		High	**Low**
Rivalry	*High*	Individual Property (finite resource)	Common Property (infrastructure)
	Low	Intellectual Property (books, music, consulting)	Knowledge Commons or Digital Commons (language, knowledge, code, free software)

Within this typology, FLOSS is positioned as a knowledge or digital commons. Digitised knowledge – in the form of source code, README files, software packages, and the shared documentation required in collaborative production – is freely available for anyone to use and at no additional cost for reproduction. One of the unique characteristics of free software as digital commons is that it avoids the free-rider problem, whereby someone who consumes or uses a resource does not give back to the community. Even if a user of FLOSS projects does not have the capability to modify code, that person can still contribute to the community simply by using the software. As an example, consider someone using the Linux-based operating system, Ubuntu. That person would not need to pay for Ubuntu or any of the software included with the operating system, but the person can still use programs and report any flaws or 'bugs' they encounter when using the software. These can be reported back to the development community so someone within the community can work on fixing the issue. Ultimately, the fix to the software can be submitted to the project manager for inclusion in a subsequent release of the software, or the fix may be distributed as an update to all users. This process is reflective of the adage 'with many eyes, all bugs are shallow' (Raymond, 2000) which makes it possible for the programs and operating system to maintain a high quality over time. In effect, the use of free software serves as a form of quality control.

Thus, free software may be positioned as a digital commons. However, there are different approaches for understanding the ontology of the commons. Antonios Broumas (2017a) offers a useful framework for understanding these differences when he identifies four different approaches: *resource-based*, *property-based*, *relational/institutional*, and *processual*. Ostrom's (1990) approach tends to position commons as *resources* or *resource systems* that are shared by a group of people, which make them susceptible to social dilemmas. In *property-based* approaches the collective property of the commons is differentiated from private and public property. *Institutional/relational* approaches attempt to account for a 'wider set of instituted social relationships between communities and resources' (Broumas, 2017a: 1509; see also Dardot and Laval, 2019). Finally, in a *processual* approach, 'commons are defined as fluid systems of social relationships and sets of social practices for governing the (re)production of, access to, and use of resources' (Broumas, 2017a: 1509). In the processual approach, commons are understood as a process or a state of becoming. This process has also been summarised by Linebaugh (2008) when he proposed the use of *commoning* as a verb, which will be discussed in greater detail in the following chapter. For the time being, however, it is worth noting my own understanding of the commons tends to fall more clearly within the *processual* or *dialectical* understanding of the commons. This approach is also nicely summarised by Broumas (2017a) when he explains the complex interaction that takes place between a producing subject and its interrelationship with an external objective environment:

> the interaction of subject and object takes the form of a subject/object, an entity that preserves certain elements of subject and object, eliminates others, and sublates the status of such an entity through the emergence of novel properties that did not exist in its generating entities (1510).

In building on this general discussion of how free software and the digital commons can be understood through different approaches, the following section will outline one of the primary threats to the commons, which is enclosure. I offer a clarification of why I have opted for a different term to describe the complex dynamics taking place between FLOSS communities and corporations.

1.4.1. Incorporation vs. Enclosure

Within certain approaches to understanding the commons – most notably the *property-based* approach – the commons are generally held in contradistinction to private property. In other words, once the commons become commodified or privatised, they cease to be commons and are in the service of capital. Even within more recent work on the revolutionary potential of the commons and commoning activities, the commons are positioned as a potential alternative to capitalism (see Dardot and Laval, 2019). The process by which commons become transformed into private property is known as enclosure. Historically, the enclosure of common land in England took place in varying degrees between the 15th century and the 19th century.[8] Enclosure took various forms throughout this period, including voluntary enclosures, forced enclosure, parliamentary legislation, and others. Throughout this process, ownership of common land was transferred to private owners, who then claimed the right to restrict access to the land. This effectively ended the open field system, whereby commoners held traditional rights to use open fields for feeding livestock, farming, or harvesting from the land. While historians still debate the extent to which enclosure exacerbated class divisions and played an integral role in the development of capitalism in general, the process nonetheless affected the relationship between commoners, capitalists, and the commonly held resources that once provided a means of subsistence for commoners. Moreover, the state played a crucial role in facilitating enclosure through the Enclosure Acts, which were passed between the 18th and 19th centuries in England and Wales (see Polanyi, 2001).

The enclosure of common land was accomplished by literally erecting fences around previously open fields. Enclosure of knowledge commons, however, depends on restricting access or prohibiting certain uses of informational resources. James Boyle (2003) refers to the process of enclosing the knowledge commons as the Second Enclosure Movement, whereby intellectual property rights restrict access to those things which were once considered common property.

Similarly, Mark Andrejevic (2007) uses the term 'digital enclosure' to refer to the process by which two distinct classes are formed online: 'those who control privatised interactive spaces (virtual or otherwise), and those who submit to particular forms of monitoring to gain access to goods, services, and conveniences' (3). In other words, Internet users, as a class, have nothing to sell but their data, which serves as a form of value production for Internet Service Providers (ISPs), which represent a class that controls the means of digital production. In this sense, the ISPs can restrict access to their sites unless users agree to the Terms of Service (ToS) or End User Licensing Agreement (EULA). These non-negotiable contracts place restrictions on how users may interact with the site. The effect of these agreements is to enclose informational resources, which are controlled by ISPs. This type of value capture has also been critiqued in debates about digital labour (see Jarrett, 2016; Fuchs, 2015; Scholz, 2013), which will be discussed further in the following chapter.

In this book, I use the term 'incorporation' rather than 'enclosure'. The term 'enclosure' implies either a physical barrier or other restriction (i.e. intellectual property rights) placed upon the commons. In effect, the 'enclosure' of digital commons typically refers to the process of imposing higher degrees of excludability on the collective resource. However, as the case studies in this book demonstrate, corporations have developed unique ways of transforming the products and processes of commons-based peer production into commercial offerings without placing restrictions on FLOSS communities' access to their common resources. This is qualitatively different from other forms of 'enclosure' discussed above. For this reason, I have opted for the term 'incorporation' because I think it more accurately describes what is happening when corporations get involved in FLOSS projects, and this will be made clear by the case studies provided in subsequent chapters. Incorporation is generally defined as the inclusion of something as part of the whole, but it also carries the specific legal definition of formally establishing an organisation as a corporation. In what follows, however, I discuss one more notable contribution for understanding the dynamics between FLOSS communities and corporations.

1.4.2. Commons-Based Peer Production

The work of Yochai Benkler (2006) is useful for understanding the broader social dynamics at work in communities of peer producers as well as how those communities intersect with existing institutions. One of the most notable contributions in this regard is his concept of *commons-based peer production* and its consequences for a broader set of social relationships. Benkler (2006) argues that commons-based peer production constitutes a new form of organisation that is 'radically decentralized, collaborative, and nonproprietary; based on sharing resources and outputs among widely distributed, loosely connected individuals who cooperate with each other without relying on either market signals or managerial commands' (60). Benkler positions social production in general and peer production specifically in contradistinction to market-based production, arguing that these forms of production constitute a form of non-market production. While these spheres are not mutually exclusive, Benkler argues that diverse forms of non-market production, like FLOSS, have the capability to influence market production.

Peer production can challenge market-based production in at least a couple of ways. First, peer production can develop products that will compete directly with those produced by commercial firms. In this case, the commercial firm has a few different options: compete, do nothing, or adopt and adapt. If the firm chooses to compete, it will be required to somehow create a better product than that offered by the nonmarket rival, although this may come at considerable cost to the firm. Alternatively, the firm can do nothing. In this case, the firm is basically relying on the belief that its products are superior to the non-market option and that the non-market option will not gain additional market share. This is a risky strategy for the commercial firm. If the non-market option does gain an increasing share of the market, the commercial firm, or at least its product that directly competes with the peer-produced option, runs the risk of becoming obsolete. The third option is to adapt to the changing forces in the market by adopting some of the strategies of the non-market forces. This type of strategic reorientation to non-market forces can have the consequence of altering the basic structure of an organisation. As Benkler (2006) notes:

> As the companies that adopt this strategic reorientation become more integrated into the peer-production process itself, the boundary of the firm becomes more porous. Participation in the discussions and governance of open source development projects creates new ambiguity as to where, in relation to what is 'inside' and 'outside' of the firm boundary, the social process is (125).

Altering the firm's position in relation to peer production, which exists outside the firm, arguably offers a higher form of risk for the firm. The firm gives up a certain level of control over the production process. The traditional view of a firm's control over its informational resources or, more specifically, knowledge, is that knowledge can be viewed as an asset to be managed as an investment (Machlup, 1962). However, the peer production process in general is far more innovative and efficient than centralised production, including outside the realm of software production (Von Hippel, 2005).

Fritz Machlup (1962) was one of the first scholars to propose that knowledge could serve as an economic resource, and his work was one of the first to popularise the idea of the information society. However, knowledge and information are typically viewed from a supply-side perspective, especially in economics literature that treats these factors as investment costs for the firm. Arguing from an alternative perspective, Frischmann (2012) suggests that we can view knowledge, information, and cultural resources as a form of intellectual infrastructure. Doing so positions these resources as 'basic inputs into a wide variety of productive activities,' which 'often produce public and social goods that generate spillovers that benefit society as a whole' (Frischmann 2012, xii). Such an argument resonates nicely with the arguments in favour of promoting commons-based peer production for enabling greater innovation (Benkler, 2006; Von Hippel, 2005). By framing knowledge and information as an infrastructural component of social development, protecting the knowledge commons becomes crucially important to the survival of commons-based peer production.

The concept of the commons is useful for thinking about informational resources. Given the increasing interconnectivity between people across vast spatial boundaries with the ability to communicate and collaborate in online environments, maintaining a base of commonly held resources that can be used for peer-production remains a central concern for facilitating more open and democratic forms of communication. This is particularly the case because the commons are subjected to the threat of enclosure or incorporation, which can threaten a community's rights of access to the commons or the collective governance of the commons.

1.4.3. Summarising Different Approaches to the Commons

The previous sections introduced the commons and commons-based peer production. Those sections drew heavily from the work of two scholars: Elinor Ostrom and Yochai Benkler. However, these scholars take different approaches to their ontological understanding of the commons. Drawing from Broumas's (2017a) framework, I positioned Ostrom's work as a *resource-based* ontology of the commons. This is because Ostrom began her analysis with the collectively governed resource, then examined the ways that communities governed those resources. The value of Ostrom's scholarship, then, was to provide a framework for understanding how communities can manage common resources outside of market relations or state provision. Rather than offering a prescriptive argument for how all communities ought to govern common resources, Ostrom's framework accounts for the diverse and varied ways that communities establish adaptable institutions of governance for managing complex problems. As such, Ostrom's project builds a 'bottom-up' approach for understanding community governance as well as the community's relationship to common-pool resources.

The work of Yochai Benkler (2006) can also be understood within the emergence of the commons paradigm, although his approach differs from Ostrom. Benkler's ontological positioning of the commons falls more within the *relational/institutional* approach, as defined by Broumas (2017a). Such an approach abstracts from simply focusing on communities or resources, and instead focuses on the social relations and structures that exist between the two. In this regard, his work focuses on the broader implications of the digital commons for economics, politics, and culture. Ultimately, he explores the greater degrees of freedom, autonomy, and creativity that are made possible by digital technologies, including the ways in which digitally networked practices of production would alter the relationship between communities and capitalist firms. In this regard, Benkler's work is also more conducive to a critical or, in Broumas's terms, a *processual* or *dialectical*, understanding of the dynamics existing between FLOSS communities and corporations.

Broumas (2017b) also offers another framework for differentiating between social democratic and critical theories of the intellectual commons that is useful in this regard. Although his framework was used to discuss the intellectual commons, the framework may also be mapped onto the digital commons. According to Broumas, social democratic theories of the commons 'employ political economic methodologies to analyse the dynamics that unfold between the commons, the market and the state with the aim to propose reconfigurations of these relations which will best serve social welfare' (103). Such theorists argue that by making progressive changes to existing structures, we can bring about a more just and egalitarian society. As it concerns the digital commons, the goal is to build repositories and platforms for commons-based knowledge and peer-to-peer production that can, in turn, bring about greater degrees of personal freedom as well as democratic decision-making (Bauwens 2005; Benkler 2006).

Table 1.4: Social Democratic and Critical Theories of the Commons (Broumas, 2017b: 121).

	Social Democratic Theories	Critical Theories
Epistemology	Political Economy	Critical Political Economy
Agency	Social Individuals	Social Intellect
Structure	Productive Community	Community of Struggle
Internal Dynamics	Bottom-Up/Top-Down Emergence	n/a
External Dynamics	Co-Existence of Commons with Capital	Commons/Capital Antagonism and Sublation
Normative Criteria	Deontological [reformist]	Deontological [subversive]
Social Change	The Commons as Substitute for the Welfare State	The Commons as Alternative to Capitalism

In the framework visualised in Table 1.4 above, Broumas (2017b) examines some of the foundational characteristics of each approach, focusing on epistemology, agency, structure, internal/external dynamics, normative criteria, and social change. Of particular interest in Table 1.4 is the relationship between the external dynamics and social change sections. The section on external dynamics in the table represents a large portion of the subsequent chapters, in which I explore the relationship between capitalism and the commons. One of the pressing questions for FLOSS specifically but for the commons more generally is whether these movements are capable of constituting alternatives to capitalism. Indeed, some recent scholarship by Massimo De Angelis (2017) specifically attempts to frame the commons as an alternative value system that is emerging from within capitalism but also one that has the potential to usher in a post-capitalist future, and this will be discussed in greater detail in the following chapter.

My goal for the next chapter is to specifically outline the contours of a critical political economy of the digital commons. To begin the transition to that task, however, the final section of this introduction discusses some of the methodology used by critical political economists in general and in this study specifically.

1.5. A Note on Methodology

The following quote from Marx (1845) comes from a section of *The German Ideology* that discusses the essence of historical materialism:

Empirical observation must in each separate instance bring out empirically, and without any mystification and speculation, the connection of
the social and political structure with production. The social structure
and the State are continually evolving out of the life-process of definite
individuals, but of individuals, not as they may appear in their own or
other people's imagination, but as they really are; i.e. as they operate,
produce materially, and hence as they work under definite material
limits, presuppositions and conditions independent of their will (Marx,
1998, 41).

The quote represents a methodological approach to inquiry that is guided by
assumptions about how reality can be understood and described. The quote
also nicely summarises the goals of researchers working within the critical
political economy of communication – that is, to connect the definite processes
of material production with broader social and political structures. Most often,
the inquiries of critical political economists of communication are directed at
large corporations that hold extensive market power and the ability to influence the production, distribution, exhibition of, or access to, communication
resources. In the process of investigation, the aim of critical political economists is to empirically investigate the material operations of corporations and
connect those operations to the broader social system. The connections made
to the social system can be situated within national boundaries while accounting for the attendant institutions (religious, legal, cultural, etc.) that encourage
or discourage certain types of behaviour, but can also be made across those
boundaries (internationally, regionally, globally).

By making these connections, political economists search for the general tendencies of capitalism rather than seeking to establish absolute laws. This allows
the inquiry to remain open to the possibility of contradictory factors, while
also allowing for an account of diverse practices both within and across media
industries. Indeed, the contradictory factors provide the illuminating moments
for critical researchers, particularly because they provide opportunity for critique and resistance. To this end, critical political economists of communication
have provided important critiques of corporations, especially the ways in which
they operate in conjunction with the general tendencies of a broader capitalist
system. As Meehan (1999) notes, 'critical scholars share an ethical obligation to
produce knowledge that accurately describes the media and reveals the hidden
dynamics whereby media corporations attempt to commercialise and control
expression in service to advertisers and ultimately to capital' (162).

To search for these 'hidden dynamics', the current study employed a *critical
interpretive* methodological approach. Maxwell (2003) describes this approach
as used by Herbert I. Schiller, a pioneering scholar working within the critical
political economy of communication tradition. When working from a critical
perspective, one situates research findings within broader bodies of knowledge
and looks for disjunctures or contradictions arising from within the field of

study. These contradictions or disjunctures can provide germane moments for research, from which previously accepted understandings can be challenged and refined. In this sense, CPEC scholars resist interpreting research findings according to their face value or as *prima facie* evidence. Rather, the research findings are brushed against the grain of alternative bodies of knowledge to situate the results within a broader set of relationships. Similarly, Mosco (2009) describes his epistemological stance as being *constitutive*. That is, critical political economy scholars resist causal, linear determinations as well as the assumption that units of analysis are fully formed wholes. Instead, critical political economists favour an epistemological position that is based on mutually constitutive processes, which act on one another throughout various stages of formation. In this sense, the approach is dialectical in that it considers both particular and more general phenomena as part of a totality of processes. These concerns are carried with the researcher throughout the research process, regardless of what type of evidence is being investigated or how it is being gathered.

To facilitate this type of investigation, critical political economists use a variety of methods. However, the selection of method is often driven by the amount of access that the researcher has to the subject being studied. When direct access to corporations is available, critical political economists rely on research methods such as interviewing, participant observation, ethnographic methods, and other methods that allow for direct observation of the life-processes of definite individuals as they operate or produce materially. In turn, these observations can be linked with the 'definite material limits, presuppositions and conditions independent of their will' (Marx 1998: 41). When we do not have direct access to corporations, critical political economists rely on documentary evidence of corporate operations and the material production taking place within the corporation. Most often, this data comes from documents that are produced by and about the corporation. To that end, the following section discusses the specific methods used in this study.

FLOSS projects depend on extensive and accurate documentation to make the development of projects run effectively and efficiently, and these documents are made publicly available so that other developers can work on the project. The source code is one form of documentation, which enables users to understand how a project works, but many FLOSS projects also contain credits files, licensing disclosures, README files, and other documents that provide essential information to users. This information, as well as the information found on publicly available discussion lists, was combined with my experiences using Linux and attending a variety of different events and meetings focused on FLOSS, including local LUG meetings and the Open Source Convention (OSCON). The aim of these documentary and first-hand experiences was to understand the dynamics between the corporations and the community of software developers, specifically how the latter negotiate their relationship with those corporations.

The advantage of researching FLOSS communities is that nearly all FLOSS projects have unique forums, bulletin boards, or wikis dedicated to providing documentation and facilitating communication about the project. These sources typically contain repositories of the project itself, but they also offer community discussion and historical data about the project's development. This, in turn, can provide documentary evidence of ongoing and past events in a way that is open to the public. For example, the Fedora Project, which is discussed in Chapter 4, features a wiki that contains extensive documentation about the project, including news, events, recent changes, user guides, and links to various sub-projects associated with the main Fedora Project. Similar sources can be found for all the FLOSS projects discussed in this study.

This introductory chapter identified the central concerns of this project by highlighting the seemingly contradictory goals of free and open source software communities and capitalist firms. Furthermore, I situated FLOSS historically by discussing some of the foundational moments in both the development of software as well as the rise of free software specifically. This discussion also included a consideration of FLOSS's cultural significance. Finally, I outlined the specific methodological approach used in the study. Now that the broad outlines and contours of the study have been established, the following chapter discusses more specifically the theoretical frameworks used to understand the complex relationships between FLOSS communities, their commons-based peer production, and capitalist accumulation.

Notes

[1] A photo of the moth that was removed from the machine is available from the Naval Historical Center at https://www.history.navy.mil/our-collections/photography/numerical-list-of-images/nhhc-series/nh-series/NH-96000/NH-96566-KN.html

[2] There is a longer history of computing research at Harvard that traces back to the 1930s, including Vannevar Bush's differential analyzer and Claude Shannon's electronic Boolean algebra. Shannon is also well known within the field of communication studies for his landmark, *A Mathematical Theory of Communication*, which was published in 1948. However, research on computing at Harvard became specifically focused on artificial intelligence in the late 1950s.

[3] The GNU Manifesto is available at http://www.gnu.org/gnu/manifesto.html (last accessed 4 January 2019).

[4] The use of the combined term 'FLOSS' is mostly pragmatic, as I am interested in exploring dynamics between the communities producing a free and/or open source software project and those corporations that sponsor or otherwise use that software. I'm interested in these dynamics regardless of whether those communities identify as free software communities, open

source communities, or some combination thereof. In certain places in the book, I specify one or the other when a distinction will be important. Otherwise, I use the FLOSS acronym for more general discussion.

[5] The supercomputer at the Oak Ridge National Laboratory is known as Summit and was built by IBM. When this computer took over the top position as the world's fastest supercomputer in June 2018, it marked the first time that a computer in the United States held that position since November 2012. In the interim, the top position was held by computers in China.

[6] The text of the GNU General Public License (GPL) can be found at http://www.gnu.org/copyleft/gpl.html (last accessed 4 January 2019).

[7] The Creative Commons Licenses can be found at http://creativecommons.org/licenses/ (last accessed 4 January 2019).

[8] A detailed account of the English enclosures is not provided here, but those interested in a more detailed treatment should see Neeson, 1993; Thompson, 1966; and Marx, 1906, especially Chapter 27: 'Expropriation of the Agricultural Population from the Land,' which is freely available at http://www.marxists.org/archive/marx/works/1867-c1/ch27.htm.

CHAPTER 2

Toward a Critical Political Economy of the Digital Commons

Existing theories of the commons come from differing epistemological stances, and they also make very different teleological propositions. Some of the more robust theorising of the commons stems from an institutional approach, which is most often associated with the work of Elinor Ostrom (1990) whose work was discussed in the previous chapter. Such an approach is valuable because it illuminates the ways in which communities cooperate to ensure the sustainability of a commons-based resource. This approach is largely descriptive and analytical in the way that it understands the commons. However, there is also a growing corpus of literature that positions the commons as an emergent value system that has the potential to either transform or replace capitalism. This approach tends to be more interpretive and prescriptive in understanding the commons and their promise for bringing about a post-capitalist future.

The purpose of this chapter is to develop a critical political economy of the digital commons that incorporates a critique of capitalism. I do so by framing the approach to this study within the critical political economy of communications tradition. Critical political economy allows for a dialectical understanding of the contradictions and tensions between capitalism and the commons. To outline these tensions, I begin with a discussion of the political economy of communications tradition. Next, I revisit the work of Karl Marx in an effort to outline the primary concerns of a critical political economy of the digital commons. Specifically, I focus on the nature of commodity production and the ways in which labour is exploited under capitalism. Then, I position FLOSS within existing debates about digital labour, while also drawing from Marxist–feminist theories of social reproduction. Following this discussion, I explore the ways the commons have been understood as an alternative to capitalism, including the ways in which the commons present an alternative circuit of value from those of capital circuits of value. As part of this discussion, I focus

How to cite this book chapter:
Birkinbine B. J. 2020. *Incorporating the Digital Commons: Corporate Involvement in Free and Open Source Software*. Pp. 33–47. London: University of Westminster Press. DOI: https://doi.org/10.16997/book39.b. License: CC-BY-NC-ND 4.0

on some of the growing critical scholarship that attempts to pair a critique of capitalism with and within theories of the commons.

Taken together, these approaches to understanding the commons are useful both analytically but, perhaps more importantly, also for the ways in which they offer proposals for a post-capitalist future. The analytical benefit, specifically as it pertains to understanding FLOSS products and processes, is that the commons paradigm can help explain how commons-based peer production and non-market production are enmeshed in processes of capitalist production. By understanding these processes more concretely, we can learn how FLOSS communities negotiate their relationship with capitalist firms and, when necessary, defend their commons-based resources from unwanted influence.

2.1. Political Economy of Communications

At the heart of the political economy of communications tradition is a concern for the 'social relations, particularly the power relations, that mutually constitute the production, distribution, and consumption of media resources' (Mosco, 2009: 24). By investigating the contours of these power relations, political economy can illuminate the ways in which power manifests itself not just as a resource to achieve goals, but also as a form of control that is embedded within a broader set of social relations. In other words, the approach allows for an understanding of power as both a preventative force (i.e. power over something else) but also as a potential force (i.e. the power to achieve change). Power relations are present throughout the social system; they structure relationships and tend to reproduce those structures over time.

To that end, those working within the political economy or, more specifically, a critical political economy of communications (CPEC), are interested in 'uncover[ing] connections between ownership, corporate structure, finance capital, and market structures to show how economics affects technologies, politics, cultures, and information' (Meehan, Mosco, and Wasko, 1993: 347). However, the concerns of those working within the CPEC tradition are not only scholarly; rather, they are often concerned with praxis or theoretically informed practice, whereby scholarly activity is pursued with the goal of achieving more just and democratic forms of communication (Mosco, 2009). Most often, this is done by exposing the ways in which power is manifested within communications industries, whereby the control of informational production, distribution, and access or exhibition is concentrated within only a handful of corporations. These large, often multinational and trans-industrial conglomerates hold oligopolistic power within media markets, which limits the possibility for alternative or counter-hegemonic forms of communication to take place (see Bagdikian, 2004; Meehan, 2005; Birkinbine, Gómez, and Wasko, 2017). By limiting the extent of available alternatives, especially by pursuing proven formulas for cultural production that generate profit for shareholders, corporations reinforce systems of ideology that, in turn, tend to reinforce institutions of cultural hegemony (Gramsci, 1971). The CPEC approach is therefore rooted in a tradition of critical inquiry, which has roots in the work of Karl Marx and his critique of classical political economy.

2.1.1. Marx, Machines, Labour, and Capitalism

By understanding FLOSS production from a critical political economic perspective, which takes inspiration from the work of Marx, we can account for the ways in which power relations structure the production, distribution, and access of informational resources. As was discussed in the Introduction to this book, FLOSS can be classified as digital commons with unique technological features – mainly, the availability of the source code and the ability to study, modify, adapt, or change the program for one's needs. However, the core value of FLOSS lies in the collective labour power of the FLOSS community. In other words, the *products* of FLOSS (i.e. the Linux kernel, Red Hat Enterprise Linux, the Fedora Project, LibreOffice, etc.) are not the source of FLOSS value, but the *processes* of FLOSS production (i.e. decentralised and distributed commons-based peer production). Because FLOSS production allows for highly efficient, collaborative, and speedy development, the end products of FLOSS production tend to be more secure, adaptable, and progressive because they are under constant revision and improvement by members of the FLOSS community. From the standpoint of corporations like Microsoft, or Oracle, which rely on the sale of proprietary software or services, FLOSS production offers an attractive option for investment because it decreases in-house labour costs and, in effect, outsources the development of core components of software that can then be integrated into their proprietary software or services. To understand the dynamics at play in cooperative production as well as the processes of commodification occurring within FLOSS production, we can revisit the work of Karl Marx.

Marx (1906) was not the first to investigate the inner workings of capitalism and the source of value within capitalism. However, he represented a shift in the study of political economy due to his criticism of previously existing political economic thought. His three volumes of *Capital* offer some of his most thoroughly developed arguments about political economy, and some of his key arguments can provide a framework for understanding the role of technology and technological change within a broader set of social relations. Although his analysis was focused on the industrial production of the mid-1800s, this background will prove useful for considering the general tendencies of capitalist production as well as the ways in which they have changed under digital capitalism.

Marx (1906) begins his analysis of capitalism with a discussion of the commodity. He explains how life appears to be an endless procession of commodities.

The commodity form, however, contains two different values: use value and exchange value. Although a commodity may contain two values simultaneously, the commodity form is still a product of human labour. That is, the process of human labour creates products in the form of commodities. Although different types of commodities require different types of labour, what is common to all commodities is human labour. The value of commodities, then, is determined by the socially necessary labour time required to produce them. These principles provide the foundation for the labour theory of value.

In early economic configurations, the trading of goods for other goods could be expressed in the simple formula: C – C (commodity for commodity trading), which characterises economies based on barter and trade. For such a trade to take place, however, the producers of such goods need to agree on an equivalence in trade (e.g.. ten apples equate to one chair). This form of trading relies on the availability of equivalent goods for such a market to operate effectively. In such a system, an apple farmer who wanted to trade apples for a chair needs certain conditions to be met to obtain the chair. First, a chair needs to be produced. Second, the chair needs to be available for trade. Third, the person who produced the chair would have a need for apples. If these criteria are met, then an exchange can occur. To reduce the uncertainty of supply and demand in such a situation, the money form (M) was introduced as a universal equivalent to which the value of all other commodities can be equated. So instead of trading ten apples for a chair, the apple farmer can sell the apples for $5. The money can then be used to buy a chair when one becomes available. The introduction of the money form, then, introduces a new type of market exchange, expressed as C – M –C (commodity for money for another commodity).

Capitalism, however, relies on larger scale production and a reinvestment in the productive process. In such a system, we can invert the C – M – C circuit to be expressed as M – C – M' , whereby money is invested in the production of a commodity with the intention of re-selling it for profit (M' or, simply, more money). This is possible in a system in which an entire class of people do not have a commodity to sell other than their labour power. In such a system, a division exists between those who own the means of production and those who do not. In other words, the owners of the means of production employ others who do not own the means of production. The engine of capitalism and the beginnings of the exploitation of labour come when the owners of the means of production only pay labourers enough to satisfy their demand, for the goal is to increase profits. By doing so, those who own the means of production continuously reinvest their money into the means of production (buying more land, developing technology, etc.). Consequently, those who own the means of production extract a certain amount of surplus value from the productive process. Thus, society is divided into classes based on ownership of the means of production (capital vs. labour).

In perhaps the most important section of *Capital*, Marx discusses surplus value in depth, including the ways in which capital continues to realise surplus value, while labour is subjected to various forms of exploitation. Particularly relevant for the current study, however, are Marx's discussions of co-operative labour and the use of machinery. Machinery is just one way in which capital constantly reinvents itself to further exploit labour. The focus on machinery is therefore simply to frame the discussion of new digital technologies and the ways that they have been used by capital and labour alike. Although technological change constantly ensures that labour is always at the mercy of capital because labour does not own the means of production, the argument presented here is that it is entirely possible for technologies to be used as tools of resistance against unwanted encroachments by capital. When put into the service of capital, technology can increase the efficiency of production and thereby increase corporate profits while further alienating labour from the production process. However, technology may be used by labour as a broader part of social resistance and social struggle.

Capital constantly seeks ways to increase surplus value, which requires more productivity by labour. This can be accomplished in at least two ways: absolute surplus labour and relative surplus labour. Absolute surplus labour is used to describe a condition in which labour is asked to work beyond the normally required working time to increase productivity. For example, workers could be asked to work through the weekend as one way of increasing productivity. On the other hand, relative surplus labour is realised when machinery supplements or supplants the time normally spent working by labour. In this sense, workers can still work the same amount of time, thereby keeping the wages owed to them constant, while human labour costs can be supplemented or supplanted by investment in a technology that performs the same function as human labour. With only limited exceptions, such a machine can be worked without the fear of fatigue or the need for sleep. Therefore, production increases without the need to pay additional wages to workers. This, then, is the key for understanding machinery (i.e. technological change) within the operation of capitalism: technology, when put in the service of capital, increases productivity, exploits labour, and is used for the realisation of greater surplus value.

Continuing this line of argument, Braverman (1974) specifically provided an extended discussion of machinery. Braverman's task was to begin a critical history of technology, which would account for the specific ways that technology has been put in the service of capital to further exploit labour. Braverman demonstrated how technological change has constantly forced labour to learn new skills to operate machinery. Furthermore, machinery has been used to supplement and supplant human labour, which drove members of the working class out of work and into unemployment. Anyone wishing to become employed again was forced to learn how to operate new machinery, which furthered the cycle of exploitation. Thus, a vicious cycle of technology development, unemployment, and re-education was implemented to constantly reinvigorate the productive process while demanding that labour constantly acquire new skills.

The relationship between capital and the labour process can also be further understood with regard to the ways that labour processes are brought under capital's control. Capitalist production is made possible by the unity of the labour process with the valorisation process (i.e. the creation and extraction of surplus value in the production of commodities). Marx uses the concepts of the formal subsumption of labour and the real subsumption of labour. The formal subsumption of labour under capital occurs when the labour process becomes subsumed under capital, whereby 'the capitalist enters the process as its conductor, its director' (Marx, 1864). In other words, the formal subsumption of labour occurs when the *social relationship* between capital and labour transforms; previously independent producers may become dependent on the capitalist through waged labour, for example. Therefore, the introduction of waged labour through becomes the social relationship between capital and labour. The real subsumption of labour occurs at a larger and more general scale when the wage labour relationship pervades social relations, thereby causing transformations within the labour process that can extract more relative surplus value. As Marx (1864) notes, 'just as the production of absolute surplus value can be regarded as the material expression of the formal subsumption of labour under capital, so the production of relative surplus value can be regarded as that of the real subsumption of labour under capital'. These concepts (i.e. absolute surplus value, relative surplus value, formal subsumption, and real subsumption) will be useful in describing the ways that FLOSS labour is exploited by capital, especially given the scale at which FLOSS projects can be developed by large numbers of geographically dispersed programmers.

Marx's analysis offers a useful framework for understanding the relationship between capital, labour, value, and machinery. These four factors are all intertwined in the relationships that exist between FLOSS programmers, their collective labour power, the software they create, and the corporations that make use of their software. The labour theory of value can be used to understand why the processes of collaborative production within FLOSS are so valuable for corporations. Collaborative production in FLOSS expands the possible labour force available to work on a software project to an exponentially greater degree than those software projects that are centralised within one firm. With more programmers contributing changes to the FLOSS software project, production and maintenance of the software can grow more efficiently and rapidly. These contributions can take the form of fixing bugs, developing new features, or increasing functionality in some other way. Because the labour of FLOSS programmers contributes to the creation of digital commons, an analysis of their labour processes can be understood within the context of theories about communication labour, digital labour, or free labour, albeit with certain distinctions.

2.1.2. Communication Labour, Digital Labour, and Its Social Reproduction

A critical understanding of capitalist production, and particularly its consequences for labour, is useful for understanding the ways that information and communication technologies (ICTs) operate today. Political economists of communication have called for increased attention to be paid to communication labourers (McKercher and Mosco, 2007; Mosco, 2006). Communication labour encompasses a wide variety of labour, including those who work directly in various media industries (i.e. television, film, music, video game, and software industries, etc.), but it also includes various types of knowledge work, digital labour, and types of free labour (McKercher and Mosco, 2007; Scholz, 2013; Lazzarato, 1996; Terranova, 2004).

The terms 'immaterial labour' and 'digital labour' have found increased currency in debates about online life. FLOSS labour can be viewed as a form of 'immaterial labour' insofar as the final products of work are 'immaterial products such as knowledge, information, communication, [or] a relationship' (Hardt and Negri 2004: 108). The term 'immaterial labour' was first introduced by Lazzarato (1996) and has since been debated by critical scholars.[9] Similar debates have occurred within critical scholarship circles about the nature of 'digital labour' (see Scholz, 2013). The primary concern in these debates has been with the nature of work and labour within the information, knowledge, and communication industries with a focus on forms of unpaid labour occurring online (see Andrejevic, 2007, 2012; Fuchs 2012). In these cases, users' online behaviours are tracked and can be transformed into an audience commodity in the same way that Dallas Smythe (1981) identified with broadcasting. Whereas Smythe argued that media programs constitute a 'free lunch' for producing audiences for advertisers, the same occurs online where companies and others seek the attention of users while data is collected about users' browsing habits. As most of us spend an increasing amount of time online during both work and non-work time, our digital labour – socially necessary time spent online – offers a more sophisticated form of the audience commodity as browsing data is extracted and transformed into value by service providers and other third-party elements (Fuchs, 2011a; McGuigan and Manzerolle, 2013; Turow, 2013).

The capture of labour value online is certainly not coincidental. Schiller (1999) frames the emergence of 'digital capitalism' within the context of neoliberal policy, which viewed digitally networked technologies as a way for expanding marketing opportunities across the globe. As such, digital technologies function merely as another way to expand capital's reach across time and space, while decreasing the amount of time necessary to send and receive information about markets. The tendency of capitalism to seek the 'annihilation of space through time' (Harvey, 1989: 205) is a familiar one, and one in which communication technologies are often employed. For example, these tendencies can be traced back to the networking of the world with telegraph cables and continues today as fibre optic cables are stretched across oceans, which provide the infrastructure for the global Internet (see Winseck and Pike, 2007; Winseck, 2017). This infrastructure provides the material basis upon which forms of digital labour and massively decentralised collaborative production can occur. This infrastructure is precisely what enables the massively decentralised and collaborative production occurring within FLOSS production.

While FLOSS production might be framed as digital or immaterial labour insofar as it is involved in the production of immaterial products like software, the exploitation of FLOSS labour occurs at two distinct points in the labour process, each of which has certain qualitative differences. On the one hand, FLOSS labour is exploited in a traditional Marxist sense of exploitation when FLOSS programmers produce software that becomes commodified by corporations. In this scenario, many (but not all) FLOSS programmers may be unpaid for their labour, meaning that the corporation selling FLOSS programs appropriates all surplus value created by the programmers. This type of unwaged labour involves the appropriation of surplus value produced by FLOSS labour in the process of producing commodities. The processes of commodifying FLOSS projects will be explored specifically in the chapter on Red Hat, as it will demonstrate how the company transformed free software into a marketable commodity that could be customised and sold to clients. On the other hand, FLOSS labour is also exploited in ways similar to other forms of digital labour like those discussed above. For example, GitHub is the largest host of software code in the world and provides one of the primary online platforms for producing software projects. In the course of producing FLOSS projects, the code for those projects may appear on GitHub. While GitHub does not directly sell data about its users, its privacy policy does indicate that 'other third parties, such as data brokers, have been known to scrape GitHub and compile data' about user activities on the site. This suggests that any FLOSS production occurring on GitHub may potentially be exploited through the appropriation of value created by online activities as a form of digital labour.

There is also a compelling question as to whether FLOSS labour is alienated from the products of its labour. Even though FLOSS labourers may make small contributions to FLOSS projects based on their unique expertise, there is a certain degree of 'ownership' – or at least a claim to stewardship of FLOSS projects – that is maintained by the community of developers over time. In fact, this is what often engenders a sense of community within FLOSS development, which is sustained over time by an association of developers who wish to see the long-term survival of their project. In this sense, if FLOSS labour can be said to be alienated from their production, it is at least qualitatively different to more classical forms of industrial production.

Of course, all FLOSS production is also dependent on the ability of FLOSS communities to reproduce themselves and their capacity to labour over time. Similarly, the object of their labour – the FLOSS project – must be reproduced over time, which requires not just the direct maintenance of the software project, but also the reproduction of the labour power of FLOSS programmers. Capitalism has always relied upon unwaged labour to ensure not only its own reproduction, but also the reproduction of the labour power of workers. In this sense, the feminist critiques of Marx that emerged in the 1970s (Dalla Costa and

James, 1975; Cox and Federici, 1976; Federici, 2012) are particularly valuable for understanding FLOSS production because they demonstrate how circuits of both capital and commons production are sustained by circuits of social reproduction. Moreover, those critiques are also useful for understanding the ways in which capital increasingly encroaches on aspects of everyday life.

The relationship between circuits of social reproduction and capital accumulation circuits can be visualised in the following way.[10] In Figure 2.1, the top line represents a simple illustration of reproduction circuits, and the bottom line represents the circuit of capital accumulation. In reproduction, money (M) obtained in exchange for labour power (LP) is used to buy commodities (C), which need to be processed by additional labour (L*). This process takes place outside of formalised working relationships (i.e. waged labour) and enables the reproduction of physical and psychological labour power (LP*), which can then be sold again to capitalists. Within FLOSS production, the cycle of unwaged reproduction can be applied in a couple of ways. First, there is a general process of social reproduction whereby FLOSS programmers reproduce their labour power over time by purchasing food, clothing, shelter, etc. and all those commodities that are required to reproduce the programmer's labour power. But there are also other ways in which the cycle of reproduction can apply to FLOSS labour. As I have already explained, a good deal of FLOSS labour is unwaged or takes place informally outside traditional forms of waged labour. Some of the specific dynamics at play here will be explored in greater detail in subsequent chapters, but one form of unwaged labour that could apply here is student labour. Money (M) could be expended by a student for additional education

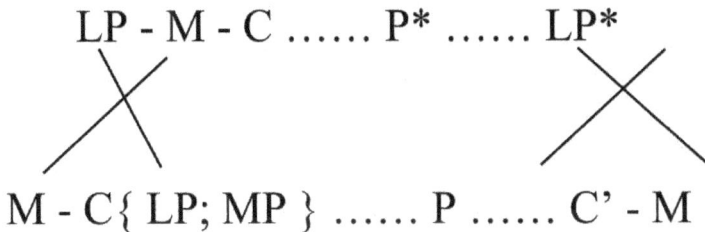

$$LP - M - C \ldots\ldots P^* \ldots\ldots LP^*$$

$$M - C\{\, LP;\, MP\,\} \ldots\ldots P \ldots\ldots C' - M$$

LP	= Labor Power
LP*	= Reproduced Labor Power
M	= Money
C	= Commodity
P	= Production
P*	= Reproduced Production
MP	= Means of Production

*Top line represents circuit of reproduction
*Bottom line is circuit of capital valorization

Figure 2.1: Coupling Between Production and Reproduction Circuits (DeAngelis, 2017: 189)

(C), which may be used to gain additional skills. These additional skills could then be used to increase the student's capacity to labour in the future (LP*).

In sum, FLOSS labour can be understood as a form of digital labour and contextualised within the rise of digital capitalism. That said, FLOSS labour has certain unique characteristics that make it more conducive, perhaps, to understand FLOSS labour as a more traditional form of labour, which was analysed by Marx. In this sense, FLOSS labour can be understood dialectically between continuity and change, whereby some of our existing understandings of labour in general continue to apply to FLOSS labour but other aspects require further elaboration. Primarily, this consideration stems from the question of whether FLOSS labour is alienated from the products of its labour in the same way that Marx described. After all, the community does maintain a certain degree of 'ownership' of their software insofar as the specific licence applied to the software allows them to retain ownership. However, even in these cases we have examples of where the wishes of the community were violated by a sponsoring corporation. Furthermore, there is also the question of the wage labour relationship between capital and labour within FLOSS communities, as not all FLOSS contributors are waged by a sponsoring corporation, but some are. This further complicates our understanding of how exploitation operates in FLOSS labour. At the very least, we may need to temper existing theories of digital labour to account for the qualitatively different ways in which labour is exploited by capital, particularly as it concerns the production, maintenance, and application of digital technologies.

2.2. Critical Theories of the Digital Commons

The preceding sections established frameworks for comprehending the ways in which FLOSS can be understood from a critical political economic perspective, including the ways in which FLOSS labour can be exploited by capital.[11] This section begins to outline the contours of a critical political economy of the digital commons. The goal of a critical political economy of the digital commons would be twofold. First, the project would illuminate the structural dynamics and power differentials that exist within commons-based communities, as well as the ways in which commons-based movements intersect with capital circuits. Second, the project would move beyond merely developing an analytical framework for understanding these power dynamics by developing a progressive political framework that could serve as a direction forward for a critical praxis of the digital commons.

As it concerned the analytical project, the previous chapter discussed different approaches for understanding the digital commons, which was aided by the frameworks developed by Broumas (2017a; 2017b). Within that framework, we positioned Ostrom within a *resource-based* understanding of the commons, and Benkler (2006) was most closely associated with the *relational/*

institutional approach. Similarly, in Broumas' (2017b) distinction between social democratic theories and critical theories of the intellectual commons, Benkler was positioned within the social democratic category. However, Benkler's work may not be so easily classified; there are times where his approach is much more conducive to a *processual* understanding of the commons. Benkler's concept of commons-based peer production contains the possibility of two useful contributions to a critical political economy. First, he discusses the ways in which commons-based peer production can alter our understanding of the relationship between communities of production and capitalist firms more generally. Second, however, commons-based peer production also focuses attention on the active production of the commons, thereby drawing attention to the labour processes involved in the creation, maintenance, and stewardship of the commons.

The analytical project of a critical political economy of the digital commons would build on the processual or dialectical understanding of the digital commons. According to Broumas (2017a), this approach frames the commons as 'fluid systems of social relationships and sets of practice for governing the (re) production of, access to, and use of resources' (1509). This definition draws attention to the social relations that are produced and reproduced alongside the relationship to the commons. Linebaugh (2008) frames this active creation by using the verb 'commoning'. In describing the practice of commoning, Linebaugh outlines four characteristics of commoning:

1) commoning is 'embedded in a particular ecology with its local husbandry';
2) it is 'embedded in a labour process' that exists in a particular field of praxis;
3) it is collective; and
4) it is 'independent of the temporality of the law and state' (44–45).

Commoning is therefore not just about understanding commons as resources but about the active pooling of common resources with a deep connection to the history, culture, and ecology of the place where they exist. As such, commoning is imbued with a complex relationship between subjectivity and the objects (i.e. common resources) to which those subjects relate. Broumas (2017a) explains that in this type of relationship 'the community itself is constantly reproduced, adapting its governance mechanisms and communal relationships in the changing environment within and outside the commons' (1509–1510).

This framework helps us to understand the commons and the complex interplay of subjectivity and community that is at work within commons-based communities. Massimo De Angelis (2017) has also developed an analytical framework for understanding how value is created and circulates within commons-based communities. He outlines this in his presentation of the commons circuit of value. This framework is also useful for understanding how commons circuits of value intersect with capital accumulation circuits.

2.2.1. Commons Circuits of Value

By combining systems theory (Luhmann, 1995), cybernetics (Maturana and Varela, 1998) and Marxist-feminist political economy (Marx 1906; Dalla Costa and James, 1975), De Angelis's task is to demonstrate how the commons can be understood as a system capable of bringing about a social revolution through ongoing iterations of commoning activity that are reproduced over time. Rather than arguing that such a revolution is imminent, however, he takes an epochal approach by focusing on how an emergent alternative value system like the commons has the potential to bring about a change in social relations. Just as capitalist social relations and subjectivities emerged in the feudal era, De Angelis views the commons as a similarly emergent value system responding to the excesses and exploitative tendencies of capitalism.

In the analytical portion of this work, De Angelis (2017) attempts to analyse the commons in the same way that Marx analysed capitalism. This leads him to develop a circuit of commons value, which accounts for the component parts of commons value systems. The circuit can be seen in Figure 2.2 below. In the circuit, an association of people (A) claims collective ownership of their commonwealth (CW), whether the sources of commonwealth are material, immaterial, commodity (C), or non-commodity (NC). This dual relationship between the association – as subjects – and their commonwealth – as objects – constitutes the commons (Cs). Then, through the activity of commoning (cm), which is derived from Linebaugh's (2008) definition of the term, the commons are reproduced over time. Framing the commons this way not only adds to a growing corpus of scholarship that makes similar claims (Dyer-Witheford, 2006; Hardt and Negri, 2009; Ryan, 2013; Gutierrez-Aguilar, 2014; Singh, 2017), but it also adds critical weight to commoning practices by demonstrating how those activities are capable of bringing about a postcapitalist future. Commoning, therefore, includes the reproduction of both the objects that comprise the commons as well as subjectivities in which mutual aid, care, trust, and conviviality are reproduced over time. For De Angelis, this commons circuit can couple with capital circuits through the commodity form. His argument is not that these two can and ought to peacefully coexist, but that they do exist.

For example, when commoners must interact with the money form of capital, they do so only as a medium of exchange to gain access to the materials

$$\begin{matrix} & A & & & A & \\ NC & \} & Cs...cm...Cs & \{ & & NC \\ c & \} CW & & & CW \{ & c \end{matrix}$$

Figure 2.2: The Commons Circuit of Value (De Angelis 2017: 193)

necessary to reproduce the commons and themselves over time. As this relates to the digital commons, a free software contributor or user still needs to have access to a computer to code the digital commons or to have access to them. In addition, the programmer will also need to have access to food, water, shelter, and all those things necessary to reproduce her own capacity to code the digital commons over time. These goods may be provided by the welfare state or one's family but, in the absence of such provision, one would need to intersect with capital circuits to obtain them. However, the extent to which commoners engage with capital circuits is left up to the community of commoners and will vary depending on the specific needs of the community.

This framework is useful for understanding the ways in which FLOSS communities relate to their digital commons. Various associations of programmers contribute to the production and maintenance of FLOSS projects, which are reproduced over time through commoning activities. The practice of commoning is a form of work that is necessary to sustain the commons over time. However, it only becomes a form of digital labour in certain circumstances. Braverman (1974), for example, draws a distinction between work and labour by explaining that work is a 'purposive action, guided by intelligence' that alters materials to improve their usefulness (49). But work becomes labour when the conception and execution of work are separated. In other words, ideas about what work is necessary can be performed by another (Braverman, 1974: 51). It is in this relationship that the division of labour occurs, which is foundational to capitalist accumulation.

At times, FLOSS communities intersect with capital circuits of accumulation when their projects are either sponsored by a corporation or a corporation incorporates a FLOSS project into their commercial offerings. As will be demonstrated in the subsequent chapters, capital exploits both the subjective qualities of FLOSS labour (e.g. collaboration, creativity, autonomy, sharing, etc.) as well as the specific objects of FLOSS labour – software that can be incorporated into commercial offerings. For example, Boltanski and Chiapello (2005) demonstrate how capitalism constantly reinvents itself by incorporating its critiques, whether they are social, aesthetic, political, or economic, into something that becomes desirable, which they refer to as the 'new spirit of capitalism.' However, despite the fact that capital attempts to encroach upon the digital commons, FLOSS communities maintain ways of negotiating and restricting access to their commonly held resources. This is particularly useful when a corporation attempts to transform the commoning activities of FLOSS programmers into labour as an input for the corporation. One of the primary means for negotiating this relationship between the community and the corporation is the establishment of 'boundary organisations'.

The concept of a 'boundary organisation' was developed within organisational theory by O'Mahony and Bechky (2008) to refer to an organisation that is set up to negotiate and establish boundaries between two parties who may have both shared and disparate interests. In effect, the organisation is established to set the terms of the relationship between two parties. Within FLOSS communities, for example, the community will want to preserve their software project while also attracting other developers to the project. The community will also want to do this while retaining rights to the software and not ceding too much control or influence to a corporation. The corporation, on the other hand, will want to use the software for commercial purposes while also asking the community to develop certain features or fix certain bugs in the software. These interests may be mutually beneficial to the community and the corporation, especially as it concerns developing effective software. However, the relationship may break down if the community feels as though the corporation is attempting to influence their activities too much. The loss of creative autonomy would almost certainly violate the norms of the FLOSS community. The specific dynamics of these types of relationships will be borne out in the subsequent chapters.

2.3. Summary

This chapter framed the study of FLOSS production within a critical political economic framework. Such an approach focuses on the ways in which corporations wield power over communication resources. Drawing from Marx's dialectical understanding of labour and capital, critical political economy focuses attention on the struggle by labour for control over communicative resources in order to bring about a more just and democratic future. As it concerns digital technology, critical political economy rejects an interpretation of digital technology as purely innovative or revolutionary, and responds by refocusing our attention on the specific cultural practices and collective labour that make up both the technology and its attendant practices.

In addition, I positioned the collective labour – or commoning activities – of FLOSS communities as the primary source of their value. This is precisely what makes FLOSS projects an attractive option for corporations because they seek to harness this labour power to supplement their overall pursuit of profit. Given these two competing circuits of value – capital accumulation circuits and circuits of commons value – there exists a tension between capitalist firms, on the one hand, and FLOSS communities on the other. Therefore, how these two forces negotiate their relationship becomes a site of struggle and contention. At times, this relationship can be mutually beneficial and can help ensure the growth, sustainability, and attractiveness of FLOSS projects. However, at other times, this relationship can break down as capitalist firms attempt to encroach on the digital commons of FLOSS communities in various ways. The following chapters provide detailed descriptions of how these dynamics have taken shape over time. I begin with an historical discussion of the Microsoft Corporation and competing models of software production. Next, I demonstrate how Red Hat, Inc. successfully harnessed the power of the free software community to build the largest and only publicly traded corporation whose business model is entirely dependent on free software. Finally, I focus on the Oracle Corporation's acquisition of Sun Microsystems, and what happens when a corporation exerts unwanted influence in FLOSS projects. Furthermore, I explain how the FLOSS community coped with that unwanted influence.

Notes

[9] For a critique of 'immaterial labour' as an analytical concept, see Sayers, 2007.

[10] This illustration and its description is adapted from DeAngelis, 2017: 189–190.

[11] Certain portions of this section appeared in an earlier article: Birkinbine, Benjamin. 2018. Commons Praxis: Toward a Critical Political Economy of the Digital Commons. *TripleC*, 16(1): 290–305. Available via open access from https://www.triple-c.at/index.php/tripleC/article/view/929

CHAPTER 3

Shifting Toward the Commons: Microsoft and Competing Models of Software Production

The Microsoft Corporation ('Microsoft' hereafter) offers perhaps the most contentious relationship with the open source community. Primarily, this is due to Microsoft's core business model, which relies on the sale of proprietary software. Through strategic partnerships, strong intellectual property protections, and a robust strategy for capturing the consumer market for personal computer (PC) sales, Microsoft grew to become one of the largest software companies in the world. At its peak, Microsoft enjoyed nearly 97% of the market share of all computing devices in the year 2000 (Tu, 2012). This was before the company was found to be in violation of the Sherman Antitrust Act by the U.S. Department of Justice (DOJ). However, the antitrust decision did little to curb Microsoft's economic growth at the turn of the twenty-first century. Rather, the company's profits continued to grow, and Microsoft still ranks as one of the largest and most dominant software companies in the world. What has changed, particularly after the antitrust ruling, is the company's relationship to the broader free and open source software community.

As mentioned in the introduction to this book, Microsoft's former Chief Executive Officer, Steve Ballmer, referred to Linux – the open source operating system – as 'a cancer' in 2001. Slightly more than eleven years later, the company opened an entire division devoted to the promotion and development of open source software. In this chapter, the history of Microsoft's chequered relationship with free and open source software (FLOSS) is charted, focusing on three specific moments that illustrate this relationship. First, the company's initial growth and its rise as one of the most dominant software companies in the world is described. During this time, the company took an adversarial approach to open source software. This includes Bill Gates' 'Open Letter to Hobbyists' in

How to cite this book chapter:
Birkinbine B. J. 2020. *Incorporating the Digital Commons: Corporate Involvement in Free and Open Source Software.* Pp. 49–72. London: University of Westminster Press. DOI: https://doi.org/10.16997/book39.c. License: CC-BY-NC-ND 4.0

which he decried the widespread culture of freely sharing software in the hobbyist community, as well as the leak of internal documents known as 'The Halloween Documents' in 1998, which clearly outline the company's views on open source software. The second section discusses the U.S. Department of Justice's investigation and, ultimately, its conviction of Microsoft for violating the Sherman Antitrust Act. Findings from the investigation and the subsequent decrees issued to the company in the wake of the conviction are provided. The final section focuses on the most recent history of Microsoft, including its Shared Source program as well as its decision to create Microsoft Open Technologies, a wholly owned subsidiary dedicated solely to promoting and developing open source software, open standards, and open technologies.

The Microsoft case study exemplifies the clash between capital and the commons in a couple of ways. First, Microsoft's relationship with the FLOSS community is indicative of the ways in which the *processes* involved in FLOSS production transformed from a seemingly antithetical means of commercial software production into an accepted form of industrial software production. Indeed, as was discussed in the Introduction and will be seen in subsequent chapters, open source software products and processes now pervade commercial software production.

Second, the other tension between capital and the commons at the heart of the Microsoft case study can be seen in the company's stance toward intellectual property and industrial software production. On the one hand, Microsoft relies upon strong intellectual property protections to exclude others from making use of its products. Those products have been produced in-house as part of Microsoft's core business model. Microsoft uses these intellectual property rights not only to protect its own works, but to threaten FLOSS projects with infringement lawsuits. It is within this context that we can view Microsoft's long history of railing against the lack of intellectual property within the FLOSS community, beginning with Bill Gates' 'Open Letter to Hobbyists' in 1976, through to Steve Ballmer's 'Linux is a cancer' claim. What changed after the DOJ antitrust ruling is that Microsoft shifted its position toward FLOSS projects in general by submitting its own licences for approval by the Open Source Initiative (OSI). The shift in Microsoft's stance toward FLOSS after the antitrust ruling represents an important moment for Microsoft, specifically, but also for the software industry in general. The shift can be understood as a humble admission that the business model upon which Microsoft relied for most of its history had been mostly usurped by a more efficient and effective model of software production – mainly, the commons-based peer production used by FLOSS developers. But it can also be understood within the broader context of the dot-com bubble burst that hit the economy at the end of the twentieth century, which coincided with many Internet-related companies' failures but also the emergence of the Web 2.0 phenomenon. It was during this time after the DOJ ruling that Microsoft not only readjusted its positioning with respect to FLOSS projects, but also attempted to become more directly involved in FLOSS projects. The company's reasons for doing so were primarily to comply with the consent decrees to which the company agreed as part of the antitrust ruling, but also because the commons-based peer production of FLOSS had proven to be a viable and effective model of software development.

As such, capital readjusted its relationship with the emergent practice of digital commons production and sought ways to harness that production for its own gains. Two bodies of theory can be used to understand Microsoft's shift toward the commons. On the one hand, the emergent craft of FLOSS production proved to be an effective and attractive model of software development, which directly contradicted the Microsoft claims that good software development was only possible with strong intellectual property rights. In other words, the labour process involved in the production of software shifted with the growth of the smaller craft community of FLOSS development. This more generalised labour process led to a massive increase in the numbers of people working on FLOSS projects. The production taking place in that community proved capable of providing a model for industrial software production. Indeed, the processes of FLOSS production outpaced Microsoft's in-house development specifically because production was open to others. On the other hand, however, this placed pressure on labour in a couple of ways. First, FLOSS production was not subject to the same limitations as corporate software production, namely the number of working hours in a day, the number of employees working on the software, etc. This was very good for the efficiency of software production. Second, however, this feature of FLOSS production also placed downward pressure on the value of labour within the software industry.

In other words, this could be described as a mix of extracting greater degrees of absolute surplus value (i.e. extending the working day) as well as relative surplus value (i.e. technological change that decreases the value of labour). The *process* here was actually a way of extracting surplus value from software production by effectively outsourcing software production to unwaged labour. The incorporation of this labour process into industrial software production also ushered in a shift in business strategies within the software and technology industries. Instead of paying workers directly for the development of software, corporations opt to invest in technologies or platforms (i.e. fixed capital) that support open source software production. This also explains some of Microsoft's more recent ventures and acquisitions, which will be discussed toward the end of this chapter.

This chapter is structured in a way that illustrates these broader points. As such, the goal of the chapter is twofold: first, to argue that the antitrust conviction in 2001 marks a critical moment in Microsoft's history that, when paired with the bursting of the dot-com bubble and the emergence of the so-called Web 2.0 phenomenon, caused a shift in Microsoft's business strategy whereby the company tried to find ways of harnessing the power of commons-based peer production or, in other words, the labour *process* of FLOSS production. Second, it demonstrates Microsoft's own contradictory history in its stance against the open sharing of ideas. In fact, many of Microsoft's most successful products have incorporated or licensed design features that were developed by others. By making these two points, the chapter shows how Microsoft's relationship with the FLOSS community can be understood as a strategic readjustment that was undertaken in response to Microsoft's declining market share while Linux-based systems were gaining market share. Although not a complete transformation of its initial stance, Microsoft's shift in its relationship to the broader FLOSS community can be described as moving from capital toward the commons.

3.1. The Rise of Microsoft 1975–1990

Microsoft was founded in 1975 after Paul Allen and Bill Gates developed the Altair BASIC interpreter. An interpreter is a computer program that directly performs functions written in a programming language. In the case of Altair BASIC, the interpreter was designed to execute functions written in the BASIC (Beginner's Allpurpose Symbolic Instruction Code) programming language so that they could be performed on the Micro Instrumentation and Telemetry Systems (MITS) Altair 8800 microcomputer. Altair BASIC became Microsoft's first product, which was distributed by MITS under contract with the newly created company. From its very beginnings, Microsoft focused on providing software solutions that could be included on hardware devices. Microsoft's business model relied on establishing contracts with hardware providers, which would allow Microsoft products to be included on hardware.

However, the company has consistently exhibited an antagonistic position toward alleged infringements on its intellectual property. The first example of such behaviour came from unauthorised copying of its original Altair BASIC interpreter. The Altair 8800 microcomputer has been credited as the device that ushered in the microcomputer revolution (Garland, 1977). It became widely popular after being featured on the cover of the January 1975 edition of Popular Electronics. From the magazine, readers could order kits for the computer, which could then be assembled by hobbyists interested in experimenting with the device. As part of the order, readers could purchase the Altair BASIC language for a fixed price. Since the Altair BASIC language could be included with orders for the Altair 8800, Altair BASIC also became widely used. However, hobbyists often made copies for friends or others to allow them to experiment with the device as well. This made Altair BASIC subject to unauthorised copying, which prompted Bill Gates to publish an 'Open Letter to Hobbyists' on 3 February 1976.[12]

In the letter, Gates noted that 'hundreds of people who are … using BASIC' have all provided positive feedback about the interpreter. However, he claims that 'most of these 'users' never bought BASIC,' as 'less than 10% of all Altair owners have bought BASIC,' and the 'amount of royalties [Gates and Allen] have received from sales to hobbyists makes the time spent of [sic] Altair BASIC worth less than $2 per hour' (Gates, 1976: 2). Gates continued by decrying the fact that most hobbyists steal software, and asked whether this is a fair practice because it ultimately prevents good software from being written. In effect, Gates was arguing that the time, labour, and resources spent on developing software ought to be returned to him in the form of fair payment for use of the software.

Gates' open letter signalled what would become a recurring theme throughout Microsoft's history: mainly, a contentious relationship with hobbyist communities of programmers, which Gates and Microsoft viewed as infringing on intellectual property rights. The open sharing and collaboration among the hobbyist community represented a threat to Microsoft's business model, which was founded on the need to protect its products by using strong intellectual property protections. Indeed, some of the responses to Gates' open letter focused more on the business strategy, especially the shortcomings of Microsoft's contractual negotiations with the hardware vendor (Hayes, 1976). However, Gates' letter is also historically significant because it was an early document in which some of the tensions between capital and the commons were spelled out. Specifically, it highlighted tensions around labour, ownership, intellectual property, and the commercialisation of software (Driscoll, 2015). In the years that followed the Altair BASIC beginnings, Microsoft pursued a course of action that sought to do exactly that. By ingratiating itself with large hardware manufacturers, Microsoft rapidly gained market share and became one of the most dominant software companies in the world.

3.1.1. MS-DOS

Microsoft's business strategy during its early years focused primarily on providing BASIC interpreters, but the company shifted its focus to operating systems in the early 1980s. From the 1980s until the mid-1990s, Microsoft relied on the Microsoft Disk Operating System, or MS-DOS, as its core commodity. MS-DOS originated in 1981 after IBM put out a request for an operating system to use on its IBM-PC line of personal computers (PC). Shortly after the initial request from IBM, Microsoft acquired the rights to 86-DOS, an operating system from Seattle Computer Products, which it renamed MS-DOS.[13] Microsoft customised the newly acquired operating system to the specifications required by IBM. In turn, Microsoft licensed use of the operating system to IBM, which IBM then included on its IBM PCs under the name PC-DOS.

Microsoft's contract with IBM was not without controversy, however. The rise of the PC was made possible by advances in integrated circuit, or microchip, technology. Microchips for the consumer market were first used commercially in calculators, which were manufactured by companies like Hewlett–Packard and Texas Instruments. As demand for higher performance calculators increased, Intel was commissioned by Busicom, a Japanese firm, to produce the first commercially available microprocessor that could receive digital data and process it according to its programmed functions. The new microprocessor was called the Intel 4004 (Nairn, 2002). However, these new chips still needed language capable of converting instructions into signals that the chip could process. This operating system came from Gary Kildall, who authored a language capable of performing such functions. Eventually, Kildall's language was transformed into the first operating system for personal computers, known as CP/M. The rights to CP/M were held by Kildall's company, Digital Research, Inc., or DRI.

Throughout the late 1970s, CP/M became the industry leader in operating systems for personal computers. When IBM announced its initial line of personal computers, the company chose Intel as the provider for microprocessors, but it also needed a supplier for the operating system. Both Microsoft and DRI were consulted about providing an operating system. The exact details about what transpired during the negotiations are a bit murky,[14] but we know that Microsoft eventually won the contract, which resulted in the acquisition of 86-DOS that was subsequently rebranded as MS-DOS. Kildall, however, would claim that MS-DOS infringed on his copyright for CP/M. Kildall confronted both Gates at Microsoft and IBM about the alleged infringement but, on advice from lawyers, decided not to sue. Instead, Kildall chose to licence CP/M to IBM for inclusion on their personal computers. When the IBM PCs were eventually released, IBM offered a choice of operating system: $240 for CP/M or $40 for DOS (Hamm and Greene, 2004). The upshot of the dramatic price difference was that Microsoft became the clear choice for consumers, and DRI was eventually purchased by Novell in 1991.

Microsoft's contract with IBM was perhaps the biggest turning point on its path to becoming the largest software company in the world. As part of Microsoft's contract, it reserved the right to sell its operating system to third-party vendors as well, which allowed the company to exploit sales of its operating system to any hardware manufacturer. Employing this strategy, Microsoft grew tremendously from 1981–1995, with an increase in annual revenues from $16 million in 1981 to more than $6 billion in 1995 (Campbell-Kelly, 2001). Although exact figures are not publicly available, some estimates suggest that MS-DOS held nearly a 90% share of the PC market (Gilbert, 1995). Although MS-DOS would continue to be produced until September 2000, Microsoft began focusing its efforts on developing an operating system with a graphical user interface (GUI). The product that it ultimately developed, Microsoft Windows, would continue Microsoft's dominance of the personal computer software industry.

3.1.2. Microsoft Windows

Operating systems featuring a GUI did not start with Microsoft. Researchers working at Xerox's Palo Alto Research Center (PARC) first developed the GUI, which was used on the Xerox Alto computer in 1973. However, Xerox did not successfully exploit the GUI commercially. Since the market for personal computers and operating systems was already dominated by IBM and Microsoft, Xerox found it difficult to focus its efforts on commercially exploiting the GUI. Consequently, Xerox invited Steve Jobs and other representatives from Apple to its PARC for access to its prototypes in exchange for a $1 million investment in Apple prior to its initial public offering (Ward, 2013). During this visit, Jobs viewed prototypes of a computer mouse used for navigation as well as the ability to move text around on the screen. From this meeting, Jobs is said to have refocused efforts at Apple toward developing a GUI operating system. However, others have argued that assigning too much causality to Jobs' single visit is an erroneous assumption, as other Apple engineers had ties to the PARC and Jobs himself made more than one visit (Pang and Marinaccio, 2000). Whatever the inspiration, Apple worked on developing a GUI operating system for its Macintosh personal computers. However, Apple was still behind IBM and Microsoft in developing applications for its operating system.

Microsoft had established itself as a leader in the market for operating systems for PCs, and had previously worked with Apple by producing the Soft-Card, a microprocessor designed to run programs designed for CP/M on the Apple II computer. As a result, Microsoft negotiated a licensing agreement for access to the Mac operating system in 1985. At this point, Microsoft was already working on Microsoft Windows, its GUI operating system, which was announced in 1983. The purpose of the licence with Apple was to allow Microsoft access to certain visual elements of the Mac operating system so Microsoft could develop applications for the Macintosh (The History of Computing Project, 2014). To ensure that such a licence was granted, Microsoft used its powerful position in the PC software market by threatening to 'cease development work on important Mac applications unless such a license was granted' (Nairn, 2002: 375). Perhaps not coincidentally, Windows version 1.0 was released in 1985, the same year that the licence was granted.

Both Microsoft and Apple then worked on GUI-based operating systems to provide easy-to-use solutions for consumers. Although neither the first Microsoft Windows release nor the Macintosh computer proved to be commercially successful, GUI-based operating systems soon allowed massive diffusion of PCs to the consumer market. Microsoft held its IPO in 1986, which earned $61 million, which the company used to invest heavily in developing its Microsoft Windows operating system. Microsoft emerged as the clear winner during this period, and the company's relationship with IBM ensured that its operating system would be installed on IBM-compatible computers. Microsoft's growth during this period was immense, as evidenced by its growth in market share to 90% by some estimates (Gilbert, 1995). This growth in market share coincided with an increase in revenues, and the Windows operating system with its GUI was the key product that fuelled the growth. However, Apple challenged Microsoft's claims to the GUI elements of Windows, claiming that Microsoft had infringed its intellectual property. This ultimately led to a copyright infringement lawsuit between the two companies.

3.1.3. *Apple Computer, Inc. vs. Microsoft Corporation*

In 1988, Apple began a copyright infringement lawsuit against Microsoft. Apple claimed that Microsoft had infringed on 189 elements of its GUI, which, when taken together, constituted a 'look and feel' of its Macintosh operating system that was protected by copyright. Apple claimed that the infringements occurred in version 2.03 and, later, 3.0 of Microsoft Windows. The lawsuit stemmed from the initial licencing agreement that was negotiated between Apple and Microsoft when Apple granted Microsoft access to its GUI for developing applications for the Mac. The resulting litigation lasted four years, but the case was interrupted by Xerox bringing a suit against Apple, whereby Xerox claimed Apple had violated its copyrights by using some of the GUI elements originally featured in its PARC operations. Xerox further claimed that Apple was guilty of unfair business practices because of its copyright claims on the GUI, which made it difficult for Xerox to license the technology to other customers. The case against Apple grew out of the meetings held between Xerox and Apple when Steve Jobs and other Apple representatives visited the Xerox PARC to see prototypes of the GUI in exchange for Xerox's ability to acquire stock prior to Apple's IPO.

Xerox's claims against Apple were ultimately dismissed, as Apple claimed that, while it may have borrowed ideas from Xerox's PARC, those ideas were not able to be protected by copyright, and Xerox ought to settle any remaining dispute with the Copyright Office (Pollack, 1990). Similarly, Apple's case against Microsoft was rejected. Of the 189 claims of copyright infringement, all but ten were dismissed. In the end, the District Court ruled in favour of Microsoft, claiming that the remaining ten claims were over *ideas* rather than *expressions* that could be protected by copyright. Furthermore, the original licensing agreement signed between Microsoft and Apple granted Microsoft the 'right to transfer individual elements or design features using its "Windows" program' (Apple Computer, Inc. v. Microsoft Corporation, 1994).

While the details of this 1994 case may not seem directly related to corporate involvement in FLOSS, it does illustrate several things about software development, intellectual property, and Microsoft. First, the case demonstrates that early software development, particularly of those features that we may take for granted today like the GUI, was not the result of rugged individuals developing the technology alone – Richard Barbrook and Andy Cameron (1995) developed a similar critique in *The Californian Ideology*. Rather, technological development is a collective and collaborative process in which the ideas of others can influence the direction of development.

Second, the case is instructive for the exploitation of intellectual property, specifically because it illustrates how original authorship can be separated from ownership (Bettig, 1992). While the idea and design for the GUI may have originated in Xerox's PARC, Xerox had not commercially exploited its designs. Through a series of licensing agreements – first between Apple and Xerox, and later, between Apple and Microsoft – the rights to the individual elements of the GUI became diffused as they were shared among peers. Microsoft was already in a strong market position to exploit the GUI through its Microsoft Windows operating system, whereas Apple relied on assistance from Microsoft for developing applications for its emerging Macintosh computer. By doing so, however, Apple gave access to its GUI operating system to Microsoft. In turn, Microsoft honoured the stipulations of its original licensing agreement with Apple, but it would later continue development of its Windows operating system by using some of the same elements that Apple had been using. Furthermore, Microsoft's alliance with major technology manufacturers ensured that its operating system would be rapidly adopted, which further solidified its market power during the 1990s.

Third, there is a great contradiction at the heart of this case when compared with the history of Microsoft. Although the company benefited from sharing ideas to develop its Windows operating system, the company relied heavily on strong intellectual property protections to exclude others from its software as it ruthlessly defended its position atop the software industry throughout the 1990s. As we will see, however, this ruthlessness is ultimately what led to investigations for antitrust violations.

3.2. Microsoft in the 1990s

Microsoft's partnership with IBM was what ultimately allowed the company to solidify its strategic position at the apex of the computer software industry. Sales of the IBM PC and its clones reached nearly 16 million by 1990, which represented nearly 84% of the market share for personal computers (Reimer, 2005). Originally, Microsoft teamed with IBM to produce the OS/2 operating system, which IBM intended to include on its PCs, but Microsoft was busy working on its Windows operating system. When Windows 3.0 was released in 1990, the relationship between IBM and Microsoft became strained to the point that the companies decided to terminate their Joint Development Agreement,[15] which specified the partnership between the two firms for working on OS/2 (TechInsider.org, 2016). Because the Windows operating system was more developed when the companies ended their relationship, Microsoft rapidly picked up market share as its operating system was included on sales of IBM-compatible PCs. In fact, it was the relationship between IBM and Microsoft that initially drew attention from the United States Federal Trade Commission (FTC) in 1990.

The investigation by the FTC was initiated because of a joint news release by IBM and Microsoft during the Comdex trade show in Las Vegas, NV, on 13

November 1989 (Wallace and Erickson, 1992). In the press release, the companies claimed that 'Microsoft would hold back features for Windows in order to help industry acceptance of the OS/2 operating system' (Wallace and Erickson, 1992: 373). The FTC was concerned that the companies were colluding to control the market for operating systems. Ultimately, the FTC investigation ended in 1993 when the commissioners were split 2–2 on whether to bring an administrative action against Microsoft. In the same year, however, the Antitrust Division of the United States Department of Justice (DOJ) picked up the investigation, which would eventually lead to Microsoft's conviction for antitrust violations. The main issues in that case, however, did not centre around Microsoft's control of the operating system market but its business practices associated primarily with its Internet browser, Internet Explorer. Around the same time that Microsoft was seeking to solidify its position atop the computer software industry, at least three concurrent technological developments and their attendant cultural practices were emerging as challengers to the production model used by Microsoft in its rise to power. These developments were the emergence of the World Wide Web, the development of graphical web browsers, and the creation of Linux. Some of the early history of Linux has already been discussed in the introduction to this book, but some key moments in the rise of the World Wide Web and web browsers are also instructive for understanding competing models of software production. Specifically, the Browser Wars mark an important moment in the competition between Microsoft's model of software production and the emergent free and open source software movement.

3.2.1. The Browser Wars

To provide some brief historical context for the Browser Wars, earlier Tim Berners-Lee and Robert Caillau authored a proposal in November of 1990 for a hypertext project called the World Wide Web, which would provide 'a way to link and access information of various kinds as a web of nodes in which the user can browse at will' (Berners-Lee and Caillau, 1990). The creation of such a project relied on server-level applications to manage the nodes stored on the server and to facilitate the display and access of those nodes with a browser. Browsers served as the application running on a user's machine that could request access to the nodes stored on the server and display those nodes to the user. Web pages would need to be created that could store textual, graphical, or other types of information that could be accessed by users. By the end of the year in 1990, models of all these components had been created, and companies began developing browsers that would allow users to access the burgeoning technology of the World Wide Web.

In 1993, the Mosaic web browser was developed by a team of researchers at the National Center for Supercomputing Applications (NCSA) at the University of Illinois at Urbana-Champaign. The browser could display graphical content on the web and, although it was not the first browser to do so, Mosaic dramatically increased the popularity of browsing the web. Prior to its creation, most of the pages on the World Wide Web had been primarily text-based. However, Mosaic's place in the history of web browsers is perhaps best illustrated by tracing the history of its ownership and, ultimately, its transformation into the open-source web browser, Mozilla Firefox.

From its beginnings at the NCSA at the University of Illinois, the Mosaic browser spawned at least two primary companies that sought to commercially exploit the browser's technology. One company was called Mosaic Communications, and the other was Spyglass. The code base for the Mosaic browser was handled by Spyglass after an agreement was signed between the company and the University of Illinois, whereby Spyglass would retain the rights to commercially exploit the code. The other company, Mosaic Communications, created the Mosaic Netscape browser. In fact, many of the employees at Mosaic Communications had worked previously on the Mosaic browser at the NCSA, although the Netscape browser was built entirely by the team at Mosaic Communications. What was truly novel about the Netscape browser, however, was that it was made freely available to the public for personal use, which was unprecedented up to that point. Moody (2001) describes the significance of this strategy:

> Along with a beta-testing program on a scale that was unprecedented, the decision to allow anyone to download copies of Netscape free had another key effect: It introduced the idea of capturing market share by giving away free software, and then generating profits in other ways from the resulting installed base. In other words, the Mosaic Netscape release signaled the first instance of the new Internet economics that have since come to dominate the software world and beyond. (187).

Indeed, the Netscape browser began to pick up market share, and the University of Illinois noticed. To resolve any additional trademark disputes with the university, Mosaic Communications changed its name to Netscape Communications and reissued its browser under the name Netscape Navigator (Moody, 2001).

Netscape Navigator quickly picked up market share from 1994–1996, reaching its peak at nearly 90% in April 1996, according to some sources (Cusumano and Yoffie, 1998). Riding this extraordinary wave of enthusiasm for Netscape, the company held its IPO in August 1995. On the day of its IPO, shares of the company began selling at $28 and reached $58.25 by the end of the day, valuing the company at nearly $3 billion after only 18 months of operation (Moody, 2001). At that point, Netscape's IPO was the largest in history. The success of Netscape was not lost on Microsoft, and the company began to focus its efforts on developing a browser to rival Netscape. This was the beginning of the first browser war.

Since Microsoft had not devoted any significant amount of time and resources to developing a web browser of its own, the company decided not to build its browser from scratch. Rather, Microsoft approached Spyglass, which held the rights to the code of the original Mosaic browser. Spyglass had been developing its own version of Mosaic, known as Spyglass Mosaic. Microsoft negotiated a licence to use the Spyglass Mosaic code base in exchange for royalty payments for each copy of the browser issued, with an annual cap of $5 million (Elstrom, 1997).[16] The resulting browser was called Internet Explorer (IE), which was based on the same foundation as Netscape. As evidence of how aggressively Microsoft pursued its new Internet strategy, Page and Lopatka (2007) note that the company only had five or six employees working in the browser department in 1995 but had more than 1,000 by 1999.

In addition to assigning more employees to the browser division, Microsoft began packaging IE with distribution of its Windows operating system. As Microsoft had nearly 90% of the market for operating systems because of its contractual relationships with Original Equipment Manufacturers (OEMs), the company was able to quickly make gains in the market for web browsers. In effect, Microsoft was giving away copies of IE for free by bundling it with its Windows operating system. To do so, the company began distributing versions of IE to OEMs by sending discs to the manufacturers, and eventually required the OEMs to install IE with Windows 95. OEMs were prohibited from 'modifying or deleting any part of Windows 95, including Internet Explorer, prior to shipment' because of a non-negotiable licensing restriction that Microsoft placed on OEMs (*United States vs. Microsoft*, 1999, see Finding 158). This restriction did not allow OEMs to ship new PCs without IE installed. The effect on the market for web browsers was almost immediate. Figure 3.1 shows the

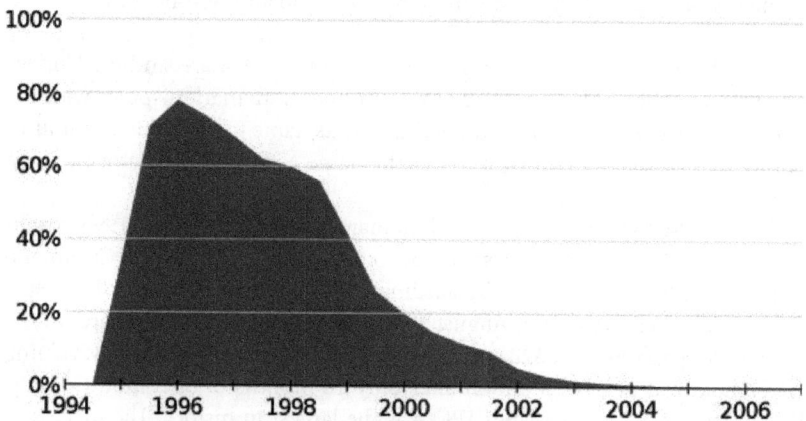

Figure 3.1: Netscape Navigator Usage Data 1994–2006 (image is in the public domain and available via Wikimedia Commons at http://commons.wikimedia. org/wiki/File:Netscape-navigator-usage-data.svg)

sharp rise in market share for the Netscape browser, and its eventual sharp decline.

Because of these tactics, Microsoft and its Internet Explorer emerged victorious in the first of the Browser Wars. Microsoft was simply too big and had too much power to influence the market for Netscape to compete. However, the novelty of distributing software freely for personal use was not lost on Microsoft. Netscape's Navigator browser rapidly picked up market share by using such a tactic, and Microsoft effectively gave away its IE browser by bundling it with its Windows operating system. Just as Microsoft was reaching its most dominant market position and using tactics that eventually led to its conviction for antitrust violations, Linux and the open-source model of production was beginning to grow as a potential threat. Indeed, after Netscape Navigator had lost significant market share to Microsoft, Netscape released the source code publicly in 1998 to attract development for a new browser. That new browser would eventually become Mozilla Firefox, which was first released in 2002. Microsoft took notice of this general trend toward open source as well and, in 1998, a series of leaked documents demonstrated exactly how Microsoft viewed this emerging threat. The Halloween Documents[17] were made publicly available and their authenticity was later confirmed by Microsoft (Harmon and Markoff, 1998). They will be discussed later in this chapter. Before doing so, however, Microsoft's conviction for antitrust violations needs to be discussed. In many ways, the antitrust conviction marks an important turning point, not just in Microsoft's history but in the broader history of the software industry.

3.3. The United States vs. Microsoft Corporation

Microsoft's activities during the Browser Wars ultimately led to its conviction for violations of Sections 1 and 2 of the Sherman Act. Section 1 of the Sherman Act prohibits 'every contract, combination ..., or conspiracy, in restraint of trade or commerce...' (15 U.S.C. §1). Section 2 states it is unlawful for any person or firm to 'monopolize ... any part of the trade or commerce among the several States, or with foreign nations. ... ' (15 U.S.C. §2). The court ultimately found Microsoft to be in violation of both sections of the Act. Microsoft violated Section 1 by unlawfully tying its web browser – Internet Explorer – to its operating system. Furthermore, the company violated Section 2 by maintaining its monopoly power by anticompetitive means and attempting to monopolise the web browser market.

These convictions rested upon the fact that Microsoft engaged in anticompetitive behaviours in its contractual relationships with Original Equipment Manufacturers (OEMs). Specifically, Microsoft used 'contractual and, later, technological shackles in order to ensure the prominent (and ultimately permanent) presence of Internet Explorer on every Windows user's PC system, and to increase the costs attendant to installing and using [Netscape] Navigator on any PCs running Windows' (United States, 2000: 11). In addition, Microsoft restricted OEMs from reconfiguring Windows 95 and Windows 98 in ways that could lead to greater use of Netscape Navigator. Finally, Microsoft 'used incentives and threats to induce' certain OEMs to design 'distributional, promotional and technical efforts' that would favour Internet Explorer instead of Navigator (*United States vs. Microsoft*, 2000: 11).

The final judgment in the antitrust case found that Microsoft had violated sections 1 and 2 of the Sherman Act, as well as more than 35 state law provisions in 19 states plus the District of Columbia. Considering these violations, the U.S. District Court Judge, Thomas Penfield Jackson, ordered Microsoft to divest its operating systems business operations from its applications business operations. In addition, all the intellectual property rights previously held by the two businesses were to be transferred to the Applications Division, which was required to grant a perpetual, royalty-free licence to the operating systems business so that it could license, develop, and distribute modified or derivative versions of the intellectual property. However, the Operating Systems Division was prohibited from doing this with the intellectual property related to the Internet browser (Internet Explorer). Aside from divesting the operations of these two businesses, Microsoft was ordered to transfer all the assets from either one of the divisions into a newly formed company, for which the transfer of ownership was to be accomplished by a distribution of stock to shareholders not connected with Microsoft. The intent of these decrees was to separate Microsoft's operating system business from the business operations that handled its web browser development. These actions would prevent Microsoft from engaging in the same types of anticompetitive behaviour that it had used during the Browser Wars.

3.3.1. Effects of the Decision

In 2001, District Judge Thomas Penfield Jackson recused himself from a related case – that went to appeal – because of some public comments that he made, which gave the impression that he had a personal bias or prejudice against Microsoft (Wilcox, 2001). In his place, U.S. District Judge Colleen Kollar-Kotelly took over the case and, in late 2001, approved a settlement between the parties. The approved settlement would no longer seek the breakup of Microsoft's Operating Systems and Applications Divisions. Instead, Microsoft agreed to a series of consent decrees in November 2002, whereby the company would be prohibited from retaliating against any OEM that develops, distributes, promotes, uses, sells, or licenses any non-Microsoft products (*United States vs. Microsoft*, 2002). In addition, Microsoft would need to establish a clearly documented schedule of all royalties that would be received from OEMs for its Windows Operating System. These provisions were aimed at prohibiting Microsoft from engaging in any anticompetitive behaviours, but most importantly for the purposes of this analysis, Microsoft would also be forced to promote interoperability for its products. This would ensure that other companies could develop products that would operate smoothly with Microsoft's products. As such, Microsoft was ordered to disclose its Application Programming Interfaces (APIs), which would specify how software components should interact with one another. By releasing its APIs to independent hardware vendors (IHV), independent software vendors (ISV), OEMs, Internet Access Providers (IAPs), and Internet Content Providers (ICP), Microsoft would ensure those parties could develop software that could operate on and interact with Microsoft's operating systems and other software. Microsoft would also need to make any communications protocol available to third parties for the same purposes. The consent decrees to which Microsoft agreed were supposed to last five years from the decision in 2002. However, these decrees were renewed twice – once in 2006 and again in 2009 – and finally expired 12 May 2011 (Chan, 2011).

In effect, the antitrust ruling against Microsoft did not seek a breakup of the company into distinct operating units, but focused more specifically on Microsoft's intellectual property practices. The decrees forced Microsoft to disclose its APIs to third parties to encourage and support interoperability with its products. The logic was that doing so would curb the anticompetitive behaviour Microsoft had displayed during the Browser Wars and in its contract bargaining with OEMs, while promoting competition within the software industry. It is within this context that Microsoft's shift toward (but not completely to) open source can be viewed.

Nevertheless, the consent decrees had little effect on the economic performance of the company. The company experienced a dip in profits in 2001, but still maintained nearly $7 billion in profits during this time with a substantial jump in the 2005–2006 fiscal year. However, along with broader shifts occurring in the software industry at the time, they did have the effect of changing some of Microsoft's practices associated with open source. The date of the consent decrees perfectly coincides with Microsoft's creation of the Shared Source program. Furthermore, the end of the consent decrees in May 2011 coincides with the creation of the Microsoft Open Technologies Division in 2012. To understand more fully Microsoft's relationship with FLOSS, the remainder of the chapter charts the company's history with FLOSS, beginning with the Halloween Documents, then discusses the Shared Source program and Microsoft Open Technologies. The previous discussion in this chapter provides an important context within which Microsoft's shift toward FLOSS can be interpreted.

3.4. The Halloween Documents

In October 1998, Eric Raymond, a well-known member of the free and open source software community and author of *The Cathedral and the Bazaar*, received a series of internal documents from a confidential source that outlined

Microsoft's strategy against Linux and open source software. These documents were subsequently released to the public by Raymond and their authenticity was later verified by Microsoft. These documents became known as 'The Halloween Documents' because many were released near the end of October over different years. The Halloween Documents focus on Microsoft's assessment of the strengths and weaknesses of open source software, including Linux, and how the company could combat the growing popularity of the movement. What is clear from the documents is that Microsoft viewed free software products as a genuine threat to its own products, especially because the free software projects had 'acquired the depth and complexity traditionally associated with commercial projects' (Raymond, 1998a). As such, the Halloween Documents contain information about how Microsoft planned to combat open source software.

In Halloween Document I,[18] Vinod Valloppillil discusses open source software as a potential threat to Microsoft. Rather than focusing on a specific open source project or organisation, however, Valloppillil focuses on the process used in open-source software development. Valloppillil writes, 'to understand how to compete against OSS [open source software], we must target a process rather than a company' (Raymond, 1998a). He goes on to assess possible strategies for combating open source software, and gives special attention to 'FUD tactics,' an acronym for Fear, Uncertainty, Doubt. FUD is a tactic used in sales, marketing, public relations, and propaganda, whereby one attempts to instil those feelings in consumers about the quality of a competitor's products. For example, in an advertisement for Microsoft Server 2003, Microsoft claimed that research demonstrated 'Linux was found to be over 10 times more expensive than Windows Server 2003' (BBC News, 2004). Microsoft was asked to change the advertisement by the Advertising Standards Authority in the United Kingdom because the results of the study were deemed to be misleading to consumers. In effect, the advertisement was meant to instil FUD in consumers about the total cost of Linux.

Halloween Document II[19] largely contains a much more detailed technical analysis of Linux's functionality when compared to other products. The author also describes his personal experience with installing the DHCP Client Daemon and ultimately claims that, even though he was a poorly skilled UNIX programmer, he could easily figure out how to extend the DHCP client code and 'the feeling was exhilarating and addictive' (Raymond, 1998b). Importantly, however, the conclusion of the document suggests possible strategies for competing against Linux. The author admits that Linux was the greatest threat to Microsoft in the server market, and he also claims that a possible strategy for fighting Linux could be patent and copyright litigation.

Halloween Document III[20] is a document from Microsoft Netherlands in which Aurelia van den Berg, a Press and Public Relations Manager for the company, responds to the leak of the two internal documents in 1998. Her response downplays the significance of the leaked documents, claiming that all companies conduct assessments of their competitors, and the leaked documents do not represent official Microsoft positions. At the end of the document, how-ever, van den Berg still manages to criticise FLOSS in general for its inability to be a long-term solution. Alluding to the need for strong intellectual property protections, van den Berg claims, 'unless Linux violates IP rights, it will fail to deliver innovation over the long run' (Raymond, 1998c).

Documents VII, VIII, and X are the other documents directly leaked from Microsoft. The remaining documents are commentaries, satires, and criticisms of Microsoft created by others in response to the leaked documents. Hallow-een Document VII[21] provides the results of an internal survey conducted by Microsoft in 2002 about attitudes and opinions on FLOSS in general, Linux specifically, and familiarity with Microsoft's newly created Shared Source pro-gram. The results of Microsoft's internal survey showed that FLOSS in gen-eral and Linux specifically were viewed favourably by those included in the survey, which mainly included policymakers, decision makers, and corporate executives selectively chosen by Microsoft. The survey also showed that mes-saging designed to criticise or question the quality of FLOSS, Linux, or the GPL was not effective (Raymond, 2002a). Considering these findings, the authors recommend that Microsoft could more effectively compete with FLOSS by focusing on the total cost of ownership (TCO) of Microsoft products when compared with Linux. In addition, the authors recommend Microsoft focus on the benefits of its newly created Shared Source program.

Halloween Document VIII[22] was an internal email sent by Orlando Ayala, Group Vice President of Microsoft's Worldwide Sales, Marketing, and Services Group, to the heads of Microsoft's subsidiaries in 2002. The message was sent as a reaction to many governments and other large institutions beginning to transition to Linux. As such, Ayala suggests that Microsoft and its subsidiar-ies need to be better prepared to respond to those types of announcements by communicating those announcements internally so the company can try to respond to these cases directly. In short, the document suggests that Microsoft's internal communication needed to be more fully integrated to respond to their declining market share, particularly among large institutions.

Finally, Halloween Document X[23] was leaked in 2004 and features an internal email from the SCO Group in which the author discusses, albeit somewhat vaguely, the relationship between the SCO Group and Microsoft. The email appears to disclose the amount of money paid to SCO on behalf of Micro-soft. Although not discussed at length here, the SCO Group was a software company that became infamous for engaging in legal battles over alleged intel-lectual property infringement in Linux related software. The SCO Group went bankrupt in 2007, but between 2003 and 2011 the company alleged that various Linux vendors had infringed copyrights belonging to it. These vendors notably included IBM, Novell, and Red Hat, but also Daimler-Chrysler and AutoZone. Particularly relevant for this discussion is the suggestion in Docu-ment X that Microsoft was contributing large amounts of money to the SCO Group to fuel intellectual property litigation against Linux and its vendors. This would be consistent with some of the suggestions in the previous documents that possible strategies for combatting Linux would be copyright and patent litigation.

In sum, the Halloween Documents allowed direct access to Microsoft's assessment of FLOSS in general and Linux specifically. What becomes clear from the documents is that Microsoft believed Linux was a legitimate threat to its own products. However, Microsoft correctly placed the true value of FLOSS projects within the process of production. To compete against the perception that FLOSS projects provided at least the same level of quality as those of proprietary companies, Microsoft used FUD tactics to suggest that the open-source model of production was inherently unstable or not secure. Ironically, Microsoft's own survey data suggested that these tactics were not effective, nor were any attempts to criticise the FLOSS development model. Instead, Micro-soft needed to shift its strategy to focus more on the quality of its own products, including its newly developed Shared Source program. The Halloween Docu-ments provide an illuminating perspective on the internal culture of Microsoft during the critical years from 1998–2004 when it underwent somewhat of a transformation. The antitrust suit against the company began in 1998 and was ultimately decided in 2001, and the company developed its Shared Source pro-gram in 2001.

3.5. Shifting Toward the Commons

The preceding sections of this chapter described in detail some of the important historical moments that exemplify competing models of software production and the specific tactics used by Microsoft to solidify its dominance of the soft-ware market. Three concurrent factors ultimately led to Microsoft's change of position regarding FLOSS. First, the company was convicted of antitrust activi-ties in 2001 and agreed to a series of consent decrees in 2002 that sought to curb the company's anticompetitive practices by requiring Microsoft to disclose its APIs to third parties. Second, the dot-com bubble burst, which marked the end of the massive speculative investment in web-based companies. Third, the rise of Linux and Linux-related businesses had demonstrated the commercial via-bility of FLOSS-based business models. Those business models – and the effec-tiveness of Linux – each relied on the *processes* involved in FLOSS production. In other words, the true source of value for FLOSS technologies and businesses was the labour performed by the FLOSS community, which provided a critical challenge to the existing models of industrial software production exemplified by Microsoft. Microsoft responded to these challenges by initiating a couple of different projects that claimed to be dedicated to FLOSS principles, although these initiatives were met with different levels of acceptance by the broader FLOSS community. The next sections chart the rise of two such projects: the Shared Source Initiative and the Microsoft Open Technologies Division.

3.5.1. Microsoft Shared Source

The Shared Source Initiative (SSI) began at Microsoft in 2001 to provide access to certain source code for debugging and reference purposes. While Microsoft had been releasing portions of its Windows source code to academic institutions and OEMs as early as 1991, the SSI expanded the range of code that was made available in 2001. The code made available under this program was protected by different licences, including the Research Source Licensing Program, Enterprise Source Licensing Program, ISV Source Licensing, OEM Source Licensing, Windows CE source code access, and others. While a detailed description of the specific rights granted by these licences and programs is beyond the scope of this analysis, these licences are mentioned here to demonstrate that the sharing of source code by Microsoft was not entirely new at the time of the antitrust ruling. However, these licences were not considered free software or open source in their true sense, because Microsoft still claimed copyright protection on the underlying source code. Under most of these licences, code was made available for academic and reference purposes, but the company prohibited redistribution of the code or limited distribution to those working on Microsoft software. In effect, these licences allowed others to view the source code, but they could modify it unless they adhered to the limitations set forth in the licences.

What was novel about the SSI in 2001 was the expansion of Microsoft's Shared Source program by the release of more types of source code as well as the creation of new licences that were designed to grant different types of rights to users. Most notable for the purpose of this project are the two licences that were submitted to the Open Source Initiative (OSI) for official registration as open source licences: the Microsoft Public License and the Microsoft Reciprocal License. Both were approved by the OSI in October of 2007 (Open Source Initiative, 2007). This marked the first time that Microsoft officially had a licence approved by the open source community, even though these licences were still not fully compatible with the GPL.

Indeed, some within the broader community viewed Microsoft's Shared Source Initiative and its new licences as simply a marketing ploy. Even Michael Tiemann, the president of OSI, the organisation that approved the licences, claimed:

> Shared source is a marketing term created and controlled by Microsoft. Shared source is not open source by another name. Shared source is an insurgent term that distracts and dilutes the Open Source message by using similar-sounding terms and offering similar-sounding promises. And to date, 'shared source' has been a marketing dud as far as Open Source is concerned. (Tiemann, 2007).

Microsoft's views differed from Tiemann's claim. In a speech in 2001, Microsoft Senior Vice President Scott Mundie noted that Microsoft's expansion of its

Shared Source Initiative may be viewed by some as a failed attempt at becoming an open source company. Mundie claimed this assertion would be false because, 'Shared Source is Open Source' (Mundie, 2001). Mundie continued by saying Microsoft would be incorporating many of the positive aspects of the FLOSS development, while continuing to preserve the company's strong intellectual property protections. Mundie went on to claim that FLOSS production was unstable as a business model in the long run because it was unsecure and subject to 'unhealthy "forking"' (Mundie, 2001). Chapter 4 will demonstrate how Mundie was incorrect in his assessment, and Chapter 5 will provide greater detail on 'forking'.

These vastly different assessments of the SSI are indicative of the contentious relationship between Microsoft and the FLOSS community. Although Microsoft had shifted its position toward FLOSS, the community still maintained a healthy scepticism about Microsoft's involvement in FLOSS projects. After all, Microsoft had a history of threatening intellectual property infringement suits against firms using Linux, even if Microsoft's stance began to thaw around the same time that Microsoft's Shared Source licences were approved by the OSI. In 2006, Microsoft agreed not to sue Novell's Linux users in exchange for a share of Novell's open source revenue, as Microsoft claimed that Novell was infringing its intellectual property. By reaching such an agreement, Novell reported that its Linux business had increased 243% through the first three quarters of the 2007 fiscal year (Lai, 2007). This agreement, as well as other similar agreements between companies using Linux and Microsoft, caused somewhat of a split within the FLOSS community as to whether companies should be signing such agreements. While the split existed in 2007, the lines of this split have blurred significantly in the years since these types of agreements began. Indeed, Microsoft opened an entire division of its company dedicated to open source, called Microsoft Open Technologies.

3.5.2. Microsoft Open Technologies and GitHub

Microsoft Open Technologies opened in 2012 to 'advance Microsoft's investment in openness including interoperability, open standards, and open source' (Foley, 2015). The creation of an entire subsidiary dedicated to open source signalled a shift in Microsoft's relationship to the broader open source community. Throughout Microsoft's history, isolated individuals or smaller working groups advocated for greater involvement in open source projects, but the creation of an entirely new subsidiary marked the first concerted institutional effort at direct involvement. Notably, the creation of the new subsidiary coincided with two major events at Microsoft. The first was the expiration of the consent decrees in 2011, and the second was the resignation of Steve Ballmer as Chief Executive Officer.

The consent decrees required Microsoft to make its APIs more openly available so that developers could create technologies that could easily interact with Microsoft's own. In other words, the consent decrees provided an impetus for forcing the promotion of greater interoperability between Microsoft and non-Microsoft technologies. In addition, Microsoft expanded its Shared Source Initiative to make its code more openly available to the broader community. However, this move was met with some scepticism by the FLOSS community, particularly because most of the licences that protected the code did not comply with open source standards. This changed in 2007 when the OSI approved two Microsoft licences as open source.

In addition to the changes brought about by the consent decrees, Microsoft experienced a change in leadership shortly after Microsoft Open Technologies opened. CEO Steve Ballmer, who is credited with the 'Linux is a cancer' indictment, announced his resignation on 23 August 2013. He ultimately resigned in 2014, and Bill Gates stepped down as Chairman of the company. However, Gates was invited to serve as technology adviser to the newly appointed CEO, Satya Nadella. Nadella adopted a new approach to open source for the company, as indicated by the actions that the company took in the years following his appointment.

In 2015, Microsoft shut down its Microsoft Open Technologies subsidiary. Microsoft did not characterise the move as closing the subsidiary but rather as Microsoft Open Technologies 'rejoining' Microsoft (Foley, 2015). The claim was that a separate subsidiary was no longer necessary, as support for open source was now mainstream within Microsoft. Indeed, a little more than a year later in 2016 Microsoft officially joined the Linux Foundation as a platinum member (The Linux Foundation, 2016). The general trend toward Microsoft's increasing support of open source was also demonstrated by the company being the top contributor to open source code projects hosted on the web-based development platform GitHub in 2017 (Hoffa, 2017). The following year, in 2018, Microsoft acquired GitHub for $7.5 billion (Microsoft, 2018).

3.6. Why Open Source? Why Now?

Microsoft's relationship with open source provides a few instructive lessons for understanding the dynamics between capital and the commons. The company's initial strategy of relying on strong intellectual property rights and enforcing them ruthlessly while simultaneously framing open source as an adversary ultimately led to an antitrust ruling shortly after the turn of the twenty-first century. Throughout the 1980s and 1990s, Microsoft's closed-source strategy and partnerships with hardware manufacturers led to its tremendous growth within the software market. The findings of the antitrust case, however, revealed the darker side of this growth. The case highlighted the company's monopolistic practices in using its dominance in the market for personal computer operating systems to distribute copies of its Internet Explorer web browser. This marked an historical turning point not just for Microsoft, but of a more general trend that saw the end of the dot-com bubble in 2001 as well as a shift away from 'Web 1.0' business tactics.

In the years after the dot-com bubble burst in 2001, a host of new web-based companies arose that promised interactivity and a focus on the consumer. This era, which marks the rise of so-called 'Web 2.0' companies, was characterised by companies providing services rather than packaged software, controlling robust data sets that expand as more people use them, trusting users as co-developers of companies' products and services, harnessing collective intelligence, relying on customer self-service, providing software across multiple devices, and featuring lightweight user interfaces, development models and business models (O'Reilly, 2005). These technological features functioned ideologically insofar as they gave the illusion of participation, collaboration, and egalitarianism when, in fact, they merely justified the provision of personal data to corporate Internet Service Providers (ISPs), who, in turn, harvested and sold that data to advertisers (see Fuchs, 2011b).

This suggests that the antitrust ruling cannot be viewed as the sole factor that affected Microsoft's business model. Rather, the antitrust decision combined with the other emerging historical forces within the technology field – Web 2.0, the commercial viability of Linux, and the ideology of romantic individualism within start-up culture – to effect a change in Microsoft's business strategy. In 2002, only a year after the antitrust ruling, Microsoft launched its 'shared source' program, which provided greater access to some of its source code, but still placed restrictions on its modification and redistribution. Consequently, the program was widely viewed as somewhat of a marketing ploy and a strategy to gain a better reputation with the open source community.

When viewed in this way, Microsoft needed to embrace open source – not only because the consent decrees required a more open approach, but because the industry in general was trending toward collaboration, and Linux (or, more accurately, the *processes* involved in FLOSS production, which made technologies like Linux possible) was proving to be commercially viable. In part, Microsoft has an interest in promoting interoperability and open standards, which enable it to keep up with the always-changing technological landscape. But the company's turn to open source may also be viewed as a humble recognition that the commons-based peer production taking place within the FLOSS community was an efficient and effective model of industrial software production that could supplement its own business practices. Finally, Microsoft's foray further into open source by its acquisition of GitHub can be understood within this broader context as well. Not only does its ownership of GitHub make the company appear as a supporter of the FLOSS community more generally, but it is also indicative of a broader trend within the information services industry of providing platforms for software production rather than directly producing software. To be sure, Microsoft does still produce proprietary software inhouse, but providing platforms for software production also places Microsoft in a strategic position that makes other forms of software production dependent on the company to a certain degree.

Microsoft remains the largest software company in the world, and it provides an example of how a corporation that was widely viewed as the antithesis to the FLOSS ethos eventually transitioned toward embracing open-source software. In effect, Microsoft is now seeking to incorporate elements of FLOSS production within its broader corporate structure. While Microsoft has not fully transformed into an open-source business, the company has shifted its position even while maintaining strong intellectual property protections over some of its core software. What is apparent, however, is that Microsoft's embracing of open source is indicative of many other large firms who are seeking to incorporate open source projects and processes into their corporate structures. Primarily, this move seems to be generated by a more general move toward cloud-based services (see Mosco, 2014). Indeed, this is further exemplified by IBM's acquisition of Red Hat, which is the largest and only publicly traded company whose business model is based entirely on free software. Exactly how the company is able to do this is the subject of the following chapter.

Notes

12 The 'Open Letter to Hobbyists' is available via the Wikimedia Commons here: https://upload.wikimedia.org/wikipedia/commons/1/14/Bill_Gates_Letter_to_Hobbyists.jpg (last accessed 4 December 2018)

13 The original name for 86-DOS was actually QDOS, which stood for 'Quick and Dirty Operating System,' but Seattle Computer Products changed the name to 86-DOS once it began marketing the product.

14 There are many different accounts of what happened. One of the most popular stories claims that Kildall snubbed the executives from IBM by choosing to go flying in his personal airplane at the time the meeting was scheduled. Other accounts claim that Kildall's wife killed the deal by insisting on changes to the contract, and others claim that Kildall did not want to release the source code for CP/M to IBM. These stories are recounted on the DRI website, which can be found at http://www.digitalresearch.biz/HISZMSD.HTM (last accessed 4 December 2018)

15 A digitised version of the Joint Development Agreement is available at https://tech-insider.org/personal-computers/research/acrobat/871126.pdf (last accessed 4 December 2018).

16 This agreement would become a point of contention between Spyglass and Microsoft, as tracking the exact number of IE copies issued proved to be incredibly difficult. Ultimately, the dispute was settled in 1997 after Microsoft agreed to issue a one-time payment of $7.5 million and an additional $500,000 in 'software and other considerations' to Spyglass (Elstrom, 1997).

17 The Halloween Documents can be found at http://www.catb.org/esr/halloween/ (last accessed 4 December 2018).

18 Halloween Document I, along with Eric Raymond's commentary, can be accessed at http://www.catb.org/esr/halloween/halloween1.html (last accessed 4 December 2018).

19 Halloween Document II, along with Eric Raymond's commentary, can be accessed at http://www.catb.org/esr/halloween/halloween2.html (last accessed 4 December 2018).

20 Halloween Document III, along with Eric Raymond's commentary, can be accessed at http://www.catb.org/esr/halloween/halloween3.html (last accessed 4 December 2018).

21 Halloween Document VII, along with Eric Raymond's commentary, can be accessed at http://www.catb.org/esr/halloween/halloween7.html (last accessed 4 December 2018).

22 Halloween Document VIII, along with Eric Raymond's commentary, can be accessed at http://www.catb.org/esr/halloween/halloween8.html (last accessed 4 December 2018).

23 Halloween Document X, along with Eric Raymond's commentary, can be accessed at http://www.catb.org/esr/halloween/halloween10.html (last accessed 4 December 2018).

CHAPTER 4

From the Commons to Capital: Red Hat, Inc. and the Incorporation of Free Software

The previous chapter focused on Microsoft's long and complicated history with free and open source software and the attendant cultural practices of open collaboration associated with FLOSS communities.[24] Microsoft underwent a transformation in its stance toward open source software. What was originally an antagonistic stance eventually transformed into an embrace of open source processes and products. In part, this was driven by the growing acceptance of free and open source software as an effective, efficient model of industrial software production, but it was also driven by the emergence of commercially viable business models that were built around FLOSS communities. Perhaps the most significant of these emergent companies was Red Hat, Inc., which became the largest and only publicly traded company whose business model was built entirely around free software.

This chapter focuses specifically on how Red Hat built its business and how it negotiated its relationship with the community of free software developers upon which its business model depends. In effect, Red Hat transformed the commons of free software production into a capitalist enterprise by trans-forming FLOSS *products* into commodities that could be customised, sold, and serviced for its customers. I understand commodification simply as the transformation of use values into exchange values, which stems from Marx's analysis of the commodity form. However, some scholars like Meretz (2014) argue that free software is not a commodity and cannot be since this is prohib-ited by the GNU General Public License (GPL). Meretz's point is that the GPL promotes direct reciprocity between people because the licence stipulates that anyone using GPL-protected works must make their subsequent work available under the same licence. On this point, I agree with Meretz. However, as I will

---

demonstrate in this chapter, Red Hat transforms the use values of free software projects into exchange value through trademark law, thereby maintaining the reciprocity of its free software projects as stipulated by the GPL while simultaneously circumventing some of those provisions by embedding its trademark into customised free software packages. In effect, this contains the hallmarks of classic commodification (i.e. the transformation of use values into exchange values) while also some elements of knowledge rent extraction when Red Hat serves as the *de facto* 'owner' of the free software commons for the purpose of market exchange.

More than any other case study, this chapter illustrates the complex ways in which a FLOSS community and its software projects can be dialectically situated between the commons and capital. After all, there are processes of commodification taking place in this example, as will be demonstrated during a discussion of Red Hat's core commodities. However, there are certain unique characteristics of those software projects that allow their code to be commodified by Red Hat, while the community continues to have access to and a certain degree of 'ownership' of the code. This relationship is mediated through the specific intellectual property licences assigned to the code in question, which will also be explored in this chapter. This is particularly notable because Red Hat continues to enjoy a favourable reputation within free software communities, and it also found a way to commodify software without enclosing or dispossessing the commons from them. Rather, the relationship between Red Hat and the free software projects that it sponsors is negotiated through what O'Mahony and Bechky (2008) call 'boundary organisations'. Such organisations are created to negotiate and establish boundaries between two parties who may have both shared and disparate interests. On the one hand, FLOSS communities want to ensure the survival of their software projects and attract other developers to work on them, which can be achieved through securing corporate sponsorship of a project. However, the community also wants to preserve its creative autonomy by not ceding too much influence or power to the corporation. Negotiating these boundaries can effectively be achieved by establishing a boundary organisation, which serves as a forum for negotiating these interests while simultaneously serving as an intermediary between FLOSS communities and corporate sponsors.

To illustrate the specific dynamics at work in the relationship between Red Hat and free software communities, this chapter first explains the history of Red Hat as well as how the company developed a way to transform the digital commons of free software into a capitalist enterprise. The specific focus is on its core commodities – previously Red Hat Linux and now Red Hat Enterprise Linux, both of which rely on collaborative commons-based peer production from within the FLOSS community. Then, the chapter focuses on the ways in which Red Hat negotiates relationships with the FLOSS community through the boundary organisation of the Fedora Project Council as well as the Contributor Licensing Agreements (CLAs). These agreements protect Red Hat against any claims to ownership by community members. Since the intellectual property rights of user contributions are centralised within Red Hat, the company then embeds its trademarked corporate logo into the distributions it sells, which gives it the ability to restrict access to and redistribution of its commodities. Finally, the chapter concludes with reflections about the Red Hat business model and what it means for the broader FLOSS community.

4.1. The Political Economy of Red Hat, Inc.

Red Hat Software, Inc. was founded in 1995 when open source software was still an emerging but rapidly growing phenomenon. In 1991, Linus Torvalds released the code for his Linux kernel project. At that time, the market for software and, more specifically, the market for operating systems was still dominated by large firms, most notably Microsoft and its Windows operating system as discussed in the previous chapter. In 1993, Bob Young formed a company, the ACC Corporation, which primarily sold Unix- and Linux-related accessories and books, and Mark Ewing created his own distribution of Linux, called Red Hat Linux, in 1994. One year later, Red Hat Software, Inc. (simply referred to as 'Red Hat' from here onwards) was founded after Bob Young's ACC Corporation merged with Mark Ewing's company. Red Hat was founded with the purpose of developing a commercially viable business model for open source by lending credibility to the emerging open source phenomenon. The creation of Red Hat was intended to bring the power of open source to businesses by providing packaged solutions to customers, while funnelling their earnings back into the open-source community by supporting free software projects. As Bob Young declared in 1999:

> We recognised the value of giving customers control of their software, and sought to bring brand reliability to the Linux product. We would offer support to customers and accelerate development of the operating system by investing our own R&D [research and development] dollars in new Linux technology that would then be given back for free to the community, for any Linux programmer or distributor to use. We had no intention of ever 'owning' the intellectual property we created. Instead, our business model was based on quickly expanding the market, and earning a small amount of revenue from a large number of customers who would buy a product that was better quality than that being offered by the industry leader, Microsoft. (Young and Rohm, 1999: 10)

The 'better quality' product that Young is referring to is the Linux-based operating system, which is created by open collaborative development, as opposed to closed proprietary development used by Microsoft. Red Hat found a way to offer an operating system that could be easily adapted to the unique needs of different customers. This was particularly important in a time when hardware vendors were reliant on large, proprietary firms such as Microsoft to develop operating systems that could run on their hardware. The speed at which new versions of proprietary operating systems could be developed was much slower compared to the open source options. Consequently, Red Hat negotiated – and continues to rely on – strategic partnerships with hardware manufacturing companies, such as Intel, IBM, Dell, Cisco, Hewlett-Packard, Sony and others.

These partnerships are beneficial to Red Hat and its partners for several reasons. First, Red Hat can pursue its original goal of bringing commercial credibility to free and open source software by gaining the support of major information technology firms. Second, Red Hat positions itself as a leading company dealing solely in free software. Third, Red Hat supports free software projects financially to support the developer communities that work on these projects. In effect, Red Hat serves as an intermediary between large information technology firms and the FLOSS community.

However, in the early years of Red Hat, the company benefited from venture capital investment, particularly at a time when the 'dot-com' investment bubble was on the rise. Frank Batten, Jr., through Landmark Communication, was an early investor in Red Hat and committed $2 million to the company in 1997 (Young and Rohm, 1999). Landmark Communication was famous for investing in the Weather Channel, and the company remains a privately held investment firm that now operates under the name Landmark Media Enterprises. Red Hat also received investment capital from Greylock Limited Partnership and Benchmark Capital, a company based in Menlo Park, CA, and known for its investment in, and support of, the open-source community. All three of these entities – Landmark Communication, Greylock and Benchmark Capital – became major shareholders in Red Hat after its initial public offering (IPO).

Red Hat held its IPO in August 1999. The investment from venture capital firms, as well as the company's partnerships with major information technology companies, led to rapid growth in the firm's value. In September 1999, Red Hat's stock price rose to more than $122 per share, up from its original price of $14 per share. At the time, Frank Batten, Jr, owned 15 million shares in the company, while Greylock Limited Partnership owned 8.7 million shares, and Benchmark Capital owned 5.8 million shares (Kanellos, 2002). However, in the interest of giving back to the FLOSS community, the company tried to compile a list of all FLOSS developers who contributed to Linux and other FLOSS projects. While arriving at a fully comprehensive list was not possible, the company managed to develop a list of approximately 5,000 developers. The intention was to make these developers stockholders in the company so they could benefit from the company's growth. While the United States Securities and Exchange Commission regulations prevented a large portion of these developers from becoming investors,[25] more than 1,000 of the eligible 1,300 developers became early shareholders in the company (Young and Rohm, 1999). Making the effort to include members of the FLOSS community as early shareholders in the company demonstrated Red Hat's commitment to supporting the community.

In the years following the IPO, Red Hat continued to enjoy growth in revenue. What is particularly striking about Red Hat's growth was that the company was not significantly affected by the dot-com bubble crash between 1999–2001. Rather, Red Hat emerged from this period and continued to grow. One reason for the company's steady growth during this period may be the strategic partnerships that Red Hat negotiated with large information technology firms in the lead up to the dot-com crash. Those firms – Intel, Cisco, IBM, Dell, etc. – also survived the crash and many have solidified their position as leaders in the market for information and communication technologies. Even though Red Hat was a start-up company, the partnerships that the company formed with these larger firms ensured that Red Hat would be supported by these businesses into the future.

While the company continued to enjoy growing revenues, its net profits exhibited a noticeable decrease during the dot-com bubble crash. Red Hat's profits dipped from 1998 until 2002, but rose again in 2003. This performance almost perfectly coincides with changes in management, and can also be explained by a shift in Red Hat's business strategy. In 1999, the original co-founders, Bob Young and Mark Ewing, left the company. In 2001, Paul Cormier joined Red Hat and began to lobby in favour of shifting the company's business model. Specifically, Cormier wanted to provide FLOSS solutions at the enterprise level rather than in the consumer market. To more fully explain the nuances of this shift, the following section contains an in-depth discussion of Red Hat's core products, how those commodities shifted focus over time, and how Red Hat centralised intellectual property within its corporate structure.

4.2. Red Hat's Core Commodities and Intellectual Property

Red Hat's business model relies primarily on its ability to provide an easy-to-use and accessible version of Linux by producing packaged distributions of the operating system, while also providing services and customer support that cater to its products. Red Hat's revenue comes from these two streams. The majority of Red Hat's revenue is derived from a subscription-based model, whereby clients get both products and support from Red Hat, in exchange for a fee. The types of products and services provided depend on the level of subscription. The effectiveness of this subscription model is based, to a large degree, on two interrelated factors: Red Hat's recognition as a trustworthy provider of FLOSS products and services, as well as Red Hat's position as a legally-recognised institution, which can be held liable for the products and services it provides.[26]

Most importantly for its customers, Red Hat provides a way to outsource services that may otherwise be too expensive to perform within a company. Indeed, any one of Red Hat's customers could perform the work done by Red

Hat, especially because the underlying code on which Red Hat relies is free software. Red Hat does not own the intellectual property rights for the free software that its services are based upon, and the company is not necessarily trying to exclude others from this intellectual property. Rather, Red Hat has built its business model on free software that is protected by the GNU General Public License (GPL), as well as other FLOSS licences. As such, any of its customers could, in theory, produce the same software that is sold by Red Hat, but they would need to perform the work themselves. However, Red Hat is liable for the products and services it supplies, which reduces the risk of in-house software development. This means that its customers can presumably be reassured that support will be available when they sign a contract with Red Hat. In effect, this is how Red Hat has become the market leader providing FLOSS distributions and services to earn revenues. Prior to Red Hat's founding, FLOSS projects had differing degrees of trustworthiness. By forming a corporate entity that could be held liable for the products and services it provided, Red Hat provided a certain degree of legitimacy to a system of production that was massively distributed and not necessarily driven by market forces. Such a system engendered projects that varied in their attractiveness to developers, which threatened the ability of certain projects to survive.

In what follows, I explain exactly how Red Hat has been able to profit from free software. I begin with a discussion of Red Hat Linux, which was the original operating system sold to customers from 1994–2004. Then, the company shifted its strategy to focus more on providing business solutions with Red Hat Enterprise Linux. Most importantly, I address the relationship between Red Hat's core commodities and the Fedora Project, which is one of the major FLOSS projects supported by Red Hat.

4.2.1. Red Hat Linux

When Red Hat first began offering products and services in the early 1990s, it sold a compact disc (for approximately $50) that contained a Linux distribution called Red Hat Linux, some additional applications and documentation. Red Hat Linux was based purely on computer code that was protected by the GPL and other FLOSS licences – that is, code that must remain freely available for distribution, modification, adaptation, etc. Red Hat Linux provided the principal source of revenue for Red Hat during its early years. Revenue came primarily from sales of Red Hat Linux to distributors and original equipment manufacturers (OEMs) for inclusion on their hardware. These companies are some of those which invested directly in Red Hat during its early years: Dell, Cisco, Hewlett-Packard, IBM and Intel. Because Red Hat had a potentially very large and distributed labour force to draw on – namely, the FLOSS community – its business model was highly scalable. That is, Red Hat had the ability to quickly expand its market share to service many customers without incurring increased investment costs. This was precisely Red Hat's strategy: to rapidly increase the market, deriving a small amount of revenue from many transactions, while reinvesting part of its earnings back into the FLOSS community.

While Red Hat Linux constituted the primary commodity for Red Hat during its early years, the bulk of its work was coming from the support it provided for this software. Red Hat's employees provided customer support, education, training and technical support to its clients. This strategy, along with Red Hat's strategic partnerships, allowed the company to pick up market share during its early years. While the company's revenues were still growing up until 2004, it had not yet become a profitable business. This was in part due to a spate of acquisitions of other software firms before the dot-com bubble crash, but also because the company had not yet found a way to substantially increase subscription sales at the enterprise level. This is precisely the change that occurred when the company shifted its focus to Red Hat Enterprise Linux, which became its core commodity and continues to be today. The final stable version of Red Hat Linux was released in 2003, which was the same year that Red Hat Enterprise Linux was released.

4.2.2. Red Hat Enterprise Linux and the Fedora Project

In 2003, Red Hat split its Red Hat Linux project into two separate projects: Red Hat Enterprise Linux and the Fedora Project. Red Hat Enterprise Linux continued as a core commodity for Red Hat in the same way that Red Hat Linux had been before. The Fedora project, however, became a community-based FLOSS project. Red Hat Enterprise Linux relied on the same model as Red Hat Linux in terms of providing packaged distributions of a free operating system but, rather than selling individual compact discs containing the software, Red Hat Enterprise Linux was made available solely through a subscription model. Depending on the level of subscription, customers could get access to customised versions of the Red Hat Enterprise Linux operating system, plus different levels of support services for it. In effect, Red Hat Enterprise Linux was a similar product to Red Hat Linux with a different customer distribution model. Red Hat then used the revenues from sales of Red Hat Enterprise Linux to support the Fedora Project. The relationship between these two projects provides perhaps the most interesting insight into how Red Hat incorporates the commons.

The split into Red Hat Enterprise Linux and the Fedora Project in 2003 was made with the intention of finding a mutually beneficial way for the FLOSS community and Red Hat to collaborate on developing software. Red Hat Enterprise Linux continues to serve as one of Red Hat's core commodities, and the company profits from subscription sales to its customers. The Fedora Project was meant to be a community-sponsored project that would provide an incubator for innovation. In return, the innovation that occurred within the Fedora

Project could then be implemented into Red Hat's commercial offerings, which could be customised to its clients' needs. This was possible because of the ownership and governance structure of the Fedora Project, as well as the worker contracts established with contributors to the project.

4.2.3. Ownership, Governance and Intellectual Property in Fedora

Red Hat, Inc. exercises ultimate control of the Fedora Project. However, the Fedora Project Council leads the Fedora Project.[27] The Council, in effect, functions as a boundary organisation for negotiating the boundaries between Red Hat and the Fedora project. However, a detailed examination of the Council is instructive for illuminating the ways in which these relationships are structured. The Fedora Project Council is comprised of six members with full voting powers: two members appointed by the community for engineering and outreach, two members elected by the community, and two members who are employees of Red Hat and are appointed by the company. The Council may also have two to four additional community members at any given time who are appointed to take the lead on specific project objectives. These members are considered auxiliary Council members with binding votes only in the areas specified by their appointment. In addition, the Council also has two additional auxiliary seats: the Diversity Advisor, who is appointed by the Council, and the Fedora Program Manager, who is appointed by Red Hat with the approval of the Council.

While the governance structure of the Fedora Project has changed over time, perhaps the most interesting factors in this structure pertain to the members appointed by Red Hat: the Fedora Project Leader and the Fedora Community Action and Impact Coordinator. The Fedora Project Leader serves as Chair of the Council, while the Action and Impact Coordinator is responsible for coordinating decision making with budgetary concerns. Previously, the Project Leader was also given veto power over any decision made by the Fedora Project Board, but now all voting members can block decisions 'with a valid reason' (The Fedora Project, 2019). However, the Project Leader does have 'a limited power to 'unstick' things if consensus genuinely can't be reached and a decision needs to be made' (The Fedora Project, 2019). The language used here is vague, but it does suggest that the Fedora Project Leader may still maintain ultimate control over the project, although he or she would presumably expend considerable political capital in making decisions that conflicted with the interests of the community.

Red Hat supports the community by sponsoring the project and directing funds to Fedora through one of its appointed employees, but it then uses the work performed by the community in its commercial offering, Red Hat Enterprise Linux. The reason Red Hat can appropriate the labour performed within the community is because all contributors to the Fedora Project have signed a contributor's agreement. These agreements have changed throughout the history of the Fedora Project, but all have similar effects. Originally, contributors needed to sign the Individual Contributor Licensing Agreement (ICLA), which effectively assigned the contributors' copyright to the Fedora Project.[28] However, the ICLA was later abandoned in favour of the Fedora Project Contributor Agreement (FPCA), which no longer assigned copyright to Red Hat, but specified the types of licences that could be included in the Fedora Project.[29] This shift made it possible for code that had already been licensed under a previous licensing scheme to be included in the Fedora Project, as long as the licences were compatible with the guidelines established by Fedora.

Both the ICLA and the FPCA provide the mechanism that allows Red Hat to commercially exploit the labour that occurs within the commons-based peer production of free software projects. In this sense, the agreements allow Red Hat to incorporate these projects into its corporate offerings by having the right to use these projects transferred to the company. In the case of the ICLA, it provided a direct assignment of a contributor's copyright to Red Hat, whereas the FPCA does not necessarily assign copyright to Red Hat. In this sense, the FPCA can be viewed as less restrictive because it allows contributors to assign licenses to their work prior to submitting the work to the Fedora Project. However, those licences must be compatible with the goals of the Fedora Project, and the Fedora Project wiki maintains a Software License List that identifies the acceptable and unacceptable licences that can and cannot be included in Fedora.[30] Importantly, Red Hat does this because it becomes legally responsible for the products that it offers to customers. If content other than code is included in the submission (text, images, logos, etc.), the contributor must waive his or her moral rights to the content. This ensures that Red Hat will not be subject to infringement claims. In effect, these licensing agreements provide a way for Red Hat to control what is included in the commons-based project (Fedora) so that when that material is included in their commercial offering (Red Hat Enterprise Linux or other software), the company will not be subject to intellectual property infringement claims by the contributors.

By taking these preventative measures to control what is included in Fedora, Red Hat can provide its customers with a guarantee that they will not need to fear a potential claim against intellectual property infringement. Red Hat does this through its Open Source Assurance Program. As the Open Source Assurance Agreement[31] contract states, if a third party alleges infringement of intellectual property in the software provided to the client by Red Hat, the company will:

> (i) defend Client against the Claim and (ii) pay costs, damages and/ or attorney's fees that are included in a final judgement against Client (without right of appeal) or in a settlement approved by Red Hat that

are attributable to Client's use of the Covered Software; (Red Hat, Inc., 2016)

Furthermore, if the Client's use of Red Hat's software is found to infringe the third party's intellectual property rights, then Red Hat will:

(i) obtain the rights necessary for Client to continue to use the Covered Software consistent with the Support Agreement(s); (ii) modify the Covered Software so that it is non-infringing; or (iii) replace the infringing portion of the Covered Software with non-infringing code of similar functionality (subsections (i), (ii) and (iii) are the 'IP Resolutions'); provided that if none of the IP Resolutions is available on a basis that Red Hat finds commercially reasonable, then Red Hat may terminate the Support Agreement(s) without further liability under this paragraph, and, if Client then returns the Covered Software that is subject to the Claim, Red Hat will refund any prepaid subscription fees related to Covered Software. (Red Hat, Inc., 2016)

From Red Hat's perspective, then, this is the legal-juridical benefit of controlling what is included in the Fedora Project, as well as centralising control of the intellectual property rights within its corporate structure. Red Hat relies on the FLOSS community to perform the cooperative labour of developing new features, fixing bugs or otherwise improving the Fedora Project so that these features can be included in its commercial offerings. To assure its customers that they will not be subject to intellectual property infringement claims from third parties, Red Hat requires contributors to assign licences to their work that will allow Red Hat to continue providing its services. In effect, Red Hat is separating authorship from ownership, which is one of the primary critiques of intellectual property laws (see Bettig, 1992). However, Red Hat does not use copyright to prevent authors or anyone else from using the code in other ways. Rather, Red Hat is trying to ensure that the rights to use the code in Fedora have been legally transferred to the company, which allows the company to provide assurances to its customers. Red Hat's method for protecting its core intellectual property does not come from copyright, but the company still prevents exact redistributions of its property through trademark law.

4.2.4. Red Hat, Trademark and CentOS

As stated earlier, Red Hat does not own the intellectual property that makes up its core commodities. Most of the code in these core commodities is covered by the GPL, which allows others to freely copy, modify and redistribute it. Therefore, rather than relying on copyright to protect its core commodities,

Red Hat relies on trademark law to protect its properties. The details of this strategy can be found in the Red Hat Trademark Guidelines[32] document (Red Hat, Inc., 2006). Hypothetically, anyone could make an exact copy of Red Hat's open source software and begin selling it, but they would be prevented from including any registered trademarks. These trademarks include the logos and names of software, which means that exact copies of Red Hat's open source software would need to be given a different name. Red Hat's trademarks also prevent products from having names that are sufficiently similar, like 'Green Hat' or 'Red Cap' or 'Redd Hatte'. While these restrictions exist, CentOS provides an example of a project that served as an exact replacement for Red Hat Enterprise Linux.

CentOS began in 2004, and served as a functionally compatible version of Red Hat Enterprise Linux. Indeed, CentOS was based on the publicly available code for Red Hat Enterprise Linux. Rather than competing with CentOS or trying to prevent them from using code included in Red Hat Enterprise Linux, Red Hat was largely ambivalent about CentOS. This was, in part, due to the perception that customers who wanted to use CentOS would probably continue to use it, but also because those customers could switch to Red Hat Enterprise Linux at any time because the two operating systems were basically the same. However, whatever tension may have existed between the two operating systems became a moot point in 2014, when Red Hat officially became a sponsor of the CentOS project. The move was perceived as a way to meet users' demands across the three major versions of Red Hat's software – Red Hat Enterprise Linux, Fedora and CentOS – by giving users access to features that may not be included across all versions of the operating system (Vaughan-Nichols, 2014). As part of Red Hat's new sponsorship of the CentOS project, all CentOS trademarks were transferred to Red Hat.

Red Hat's use of trademark law to protect its market position is deployed in conjunction with its ability to control the intellectual property included in its commercial offerings. By sponsoring the CentOS project, Red Hat can increase its intellectual property holdings, while also eliminating a rival form of free software that was offering a functional equivalent of its commercial software. In this sense, Red Hat's sponsorship of the CentOS project functions similarly to a corporate acquisition or an instance of horizontal integration.

4.2.5. Core Commodity Conclusions

Red Hat, as an institution, may be viewed in at least two different ways. On the one hand, Red Hat can be viewed as a pragmatic way to centralise commons-based peer production within capitalism. In this way, Red Hat serves as an intermediary institution for providing commercial access to commons-based peer production. In other words, Red Hat is situated between capital and the commons. Importantly, however, Red Hat is clear about its intentions and involvement in FLOSS projects, and it is one of the largest contributors to other FLOSS projects; furthermore, the company is actively paying its employees to contribute to other FLOSS projects. For these reasons, Red Hat maintains a relatively good relationship with its FLOSS communities. Indeed, Red Hat's entire business model was founded on finding a way to bring the power of FLOSS production to other businesses. In return, Red Hat reinvests in the FLOSS community by supporting FLOSS projects, acquiring new businesses and then releasing source code to the community. The relationship between Red Hat and the FLOSS community is one of mutual benefit: Red Hat's financial success benefits the FLOSS community, more revenue for Red Hat means more investment in FLOSS projects, and more investment in FLOSS projects means higher quality products and services that Red Hat can offer to its customers.

On the other hand, Red Hat can also be viewed as an institution that operates no differently to other corporations within a market-driven capitalist economy. Red Hat relies on centralising production within its corporate structure, separating authorship from ownership through worker agreements, and protecting intellectual property through trademark laws for making a profit. The difference is that Red Hat cannot prevent some actions that are commonly copyright violations because of the rights granted by free software projects. In this sense, Meretz (2014) or others who claim that free software cannot be a commodity because this is prevented by the GPL are correct, but the Red Hat case study illustrates how a company can circumvent traditional copyright law and rely on other forms of intellectual property like trademarks to become the *de facto* 'owner' of the free software commons for the purpose of market exchange. The term 'owner' is placed in quotes here because Red Hat of course is not the actual 'owner' of the commons in the traditional sense of property. However, its embedding of its trademark does allow Red Hat to, in effect, extract knowledge rent from selling customised versions of free software to its customers.

Furthermore, Red Hat does not directly employ its entire labour force, which exempts the company from directly compensating all its labourers through wages and benefits. Aside from those members of the Fedora Council that it directly employs, it relies on other informal ways of compensating those programmers who contribute to Fedora. So there is a mix of both waged and unwaged labour occurring in the production of Red Hat and the Fedora Project. In other terms, there is someting of the formal subsumption of labour (i.e. introduction of waged labour into FLOSS production), but there is also a broader point to be made about the real subsumption of labour here, because the survival of the Fedora Project is in part based upon its dependence on Red Hat. However, Red Hat relies on the development of an active Fedora community, and it is in the company's best interest to maintain a good relationship with that community. If the company were to exercise unwanted influence in the Fedora Project, those who contribute to the project may choose to abandon the project, thus ceasing development of new and innovative features that could potentially be included in Red Hat Enterprise Linux. Indeed, the following chapter illustrates what can happen when such a relationship breaks down.

4.3. From the Commons to Capital

In weighing these two interpretations, at the very least, Red Hat provides an exemplary case for understanding how the boundaries of a firm can become blurred as it orients itself toward commons-based peer production. In this sense, Red Hat demonstrates the ambiguity of commons, particularly as it pertains to the potential for radical change. Furthermore, Red Hat demonstrates how a distributed system of commons-based peer production can be centralised or incorporated into a corporation's broader strategy and turned into a profitable business. As demonstrated throughout this chapter, Red Hat accomplished this through both formal and informal mechanisms.

Red Hat was one of the earliest companies to position itself as the leading company providing services for FLOSS. As such, Red Hat sought to lend an element of professionalism to the emerging FLOSS phenomenon by establishing the formally recognised institution of a publicly traded corporation that could be legally liable for the services provided. Consequently, Red Hat needed a formalised way to control the commons-based peer production that it incorporated into its core commodities. The company accomplished this through the Individual Contributor License Agreement (ICLA) and later the Fedora Contributor License Agreement (FCLA) that granted the company rights to use the production that was performed by developers.

The contributor licensing agreements constitute a formal mechanism for controlling the informal production that takes place in commons-based peer production. These agreements are essential to Red Hat's business model because they allow Red Hat to be legally liable for the products it sells, particularly when it comes to allegations of intellectual property infringement. Red Hat is certainly not alone in using these types of agreements. The issuing of contributor licensing agreements is common practice in FLOSS projects, although the terms of the agreements may differ from organisation to organisation. Some CLAs, like the ICLA formerly used by Red Hat, represent the most striking examples of how institutions, whether for-profit or non-profit corporations, or any other type of legally recognisable organisation, formally control commons-based peer production by separating authorship from ownership. However, other CLAs like the FPCA now used by Red Hat do not require full copyright transfer. Nonetheless, CLAs in general provide a mechanism for transferring rights from commons-based peer production to commercial firms like Red Hat.

While this may be viewed as a pragmatic solution for monetising FLOSS production and products, it also illustrates the limits of Benkler's claim that the boundaries of the firm will become porous. Indeed, despite the seemingly revolutionary potential of this new modality of production, it still maintains the hallmarks of capitalist production: centralisation, control, and appropriation of surplus value. Insofar as one claims FLOSS production to be exemplary of commons-based peer production or 'non-market production', the labour performed under these conditions can still be appropriated for corporate gain. In the case of Red Hat, the company has been able to benefit from the creative input of the FLOSS community contributing to the Fedora Project. However, in the same way that Red Hat relies on both formal and informal degrees of controlling production within the Fedora Project, the company similarly relies on both formal and informal mechanisms for compensating those who contribute to its FLOSS projects.

Red Hat provides direct compensation to those members of the Fedora Council who are employed by and appointed to the Council by the company. Red Hat also directs funding back to the Fedora Project through the Open Source and Standards group, which provides funding for one of the full-time employees who serves on the Fedora Council. For those contributors who are not directly employed or paid by Red Hat, their compensation comes to them informally. Typically, community members do not have access to the budgetary funding provided by Red Hat, although community members may be elected or appointed to the Council, in which case they will at least have a say in how funds are directed. Aside from this, they may also attend public events or trade shows where institutions like Red Hat provide sponsorship or other goods and services for the community. However, this informal economy is only sustainable for as long as the institutions supporting FLOSS projects remain transparent about their intentions for the products of FLOSS developers' labour and continue to support the community through the provision of paid employment, sponsorship of additional FLOSS projects and events, and informally through gifts given to the community.

In sum, Red Hat complicates binary distinctions between market-driven production and commons-based peer production by illustrating the way that one firm has been able to implement a hybridised model of commons-based market production. Furthermore, the case study of Red Hat illuminates the contours of the ways in which the boundaries of a firm can become more porous, as was claimed by Benkler (2006). However, those boundaries are still discernible, and the production within Red Hat's corporate structure is still largely market-driven. But Red Hat, through its sponsorship of, and relationship with the Fedora Project, has found a way to move somewhat informal production from the commons to capital.

4.4. The Future of Red Hat

The preceding discussion offered a description of the way that Red Hat was able to harness free software production and transform it into a profitable business. Red Hat's attempts to include free software developers in its original IPO, as well as its ongoing contributions to FLOSS communities, earned the company a favourable reputation within programmer communities. Red Hat's ability to preserve this good reputation will be dependent, in part, on maintaining a good relationship with the Fedora Project community and not attempting to exert unwanted influence in the community.

The need for preserving this relationship has become even more urgent because Red Hat has been acquired by IBM (Red Hat, 2018). This news was announced shortly before this manuscript was submitted to the publisher. While it is still too early to tell the consequences of the acquisition, especially for the Fedora Project, I wanted to add a coda to this chapter to address the acquisition. While any prognostications for what will become of Fedora are purely speculative, there are certain factors that suggest the Fedora Project is likely to survive, even if the institutional arrangements between Fedora and IBM are altered slightly from the institutional arrangements between Fedora and Red Hat. First, and perhaps most significant, is the fact that IBM has also been supporting various FLOSS projects throughout its history, and the company is likely to respect the boundaries of the Fedora Project and its creative autonomy. Second, we have already seen examples of what can happen when a company exerts unwanted influence over FLOSS projects, which is precisely the subject of the following chapter. In that case, Oracle acquired Sun Microsystems, which had been supporting various FLOSS projects. After Oracle interfered in those projects, the communities abandoned them, leaving Oracle without any developers working on the projects. This is one of the risks that IBM will take if it decides to meddle in the Fedora Project in the wake of its acquisition of Red Hat.

Notes

[24] An earlier version of this chapter appeared as Benjamin Birkinbine, 2017. From the Commons to Capital: Red Hat, Inc. and the Business of Free Software. *Journal of Peer Production* 10. Accessed 2 January 2019. http://peerproduction.net/issues/issue-10-peer-production-and-work/from-the-commons-to-capital/

[25] Two regulations are most significant here. First, you must be a US-based taxpayer to buy IPO shares for a company listed on an American exchange. This regulation eliminated approximately half of the eligible investors. Second, since the SEC designates IPO offers as extremely high-risk investments, it regulates against 'inexperienced investors' buying shares in IPOs. This regulation eliminated another 15% of developers, as they were either students or qualified as 'inexperienced investors' according to SEC guidelines.

[26] Unless otherwise noted, the information in this section comes from Annual Reports (Form 10-K, Red Hat 2000–2018) filed with the Securities and Exchange Commission in the United States between the years 2000–2018.

27 Information about the Fedora Project Council is publicly available on the project's wiki, which is available at: http://fedoraproject.org/wiki/Council (last accessed on 2 January 2019).

28 Information about the Individual Contributor Licensing Agreement can be found on the project's wiki at: http://fedoraproject.org/wiki/Legal:Licenses/CLA (accessed on 2 January 2019).

29 Information about the Fedora Project Contributor Agreement can be found on the project's wiki (Fedora Project 2019) at: http://fedoraproject.org/wiki/Legal:Fedora_Project_Contributor_Agreement (accessed on 2 January 2019).

30 The Software License List can be found at: http://fedoraproject.org/wiki/Licensing:Main?%20rd=Licensing#Software_License_List (accessed on 2 January 2019).

31 The full text of the Open Source Assurance Agreement can be found at: http://www.redhat.com/legal/open_source_assurance_agreement.html (accessed on 2 January 2019).

32 The Red Hat Trademark Guidelines (Red Hat 2006) are available at: http://www.redhat.com/f/pdf/corp/RH-3573_284204_TM_Gd.pdf (accessed on 2 January 2019).

CHAPTER 5

Resisting Incorporation and Reclaiming the Commons: The Case of Oracle and Sun Microsystems

The previous two chapters focused on case studies of Microsoft and Red Hat, and discussed the ways in which the *processes* and *products* of FLOSS production became incorporated into capitalist production.[33] The chapter on Microsoft demonstrated how the company initially built its business model on strong protection of its intellectual property and fended off challenges from the emergent open-source models that proved to be an effective and efficient model of software production. Microsoft eventually shifted to embrace open source, albeit only in certain limited ways. The chapter on Red Hat demonstrated how free software could be transformed into a profitable business model by harnessing the labour power of the free software community and transforming its productive activity into commodities that could be customised, sold, and serviced for its customers. Furthermore, the chapter focused on the specific ways in which Red Hat negotiated its relationship with its free software project, Fedora, through the boundary organisation of the Fedora Project Council. This chapter will look at how a community of FLOSS developers deals with unwanted corporate encroachment into its community governance model. In other words, this chapter focuses on the *politics* involved in negotiating the boundaries between FLOSS communities and corporations. The focus on politics here is not only concerned with the governance structures in place for negotiating boundaries between the corporation and the FLOSS community, as was discussed in the previous chapter. Rather, the focus on politics here also specifically investigates the ways in which FLOSS communities can assert their interests against unwanted corporate attempts to influence production within the community. As such, politics here has the dual meaning of collective action as well as an ethical horizon toward which collective action can be directed.

How to cite this book chapter:
Birkinbine B. J. 2020. *Incorporating the Digital Commons: Corporate Involvement in Free and Open Source Software*. Pp. 89–100. London: University of Westminster Press. DOI: https://doi.org/10.16997/book39.e. License: CC-BY-NC-ND 4.0

This framing of politics, then, focuses on both the moral economy (Thompson, 1971) of the FLOSS community but also the specific tactics used in resisting unwanted corporate influence.

To do so, I focus on one of the largest software companies in the world, the Oracle Corporation (simply 'Oracle' hereafter), and its acquisition of Sun Microsystems (simply 'Sun' hereafter). Whereas Sun maintained a good relationship with the open source community by sponsoring various projects and allowing those projects to enjoy relative creative autonomy, those relations became strained after Oracle acquired Sun in 2010. After the acquisition, Oracle used a different strategy toward Sun's open source projects. In certain cases, Oracle ended open source activities, in others it tried to influence open source development to meet its own goals, and in others again it altered the way that the project was governed. In response, the community employed different strategies to protect their commons-based resources.

In this chapter, I focus on the histories of three such projects: the OpenSolaris operating system, the MySQL relational database management system, and the OpenOffice productivity software that was designed as an alternative to Microsoft Office. Throughout the chapter, I focus on the ways that the FLOSS community maintains a unique ability to leverage its collective labour power against corporate encroachment into its projects by using technical, legal, and governance strategies that allow them to abandon a project without losing the products of their labour. This has a similar effect to a factory walk-out, whereby workers halt the productive process by abandoning the site of production. When dealing with software, however, production is not reliant on a particular space. Rather, productive activity can simply be moved to a new location. And, because of the unique legal institutions and technical features of open source software, a project can be 'forked' whereby the project can be copied and production can continue under a new name without violating the intellectual property protections of the original project. As we will see, this is one of the primary ways that the FLOSS community leverages its collective labour power against undue corporate influence.

5.1. The Oracle Corporation and Sun Microsystems

Oracle Corporation is one of the largest software companies in the world. The company has three main operating segments: cloud and licence business, hardware, and services.[34] From these Oracle earns approximately 82% of its total revenue from the cloud and licence business segment. In 2018 alone, the company earned more than $39 billion in total revenues and employed approximately 137,000 people. If calculated by total revenues, Oracle is the third largest company in the global software market behind only IBM and Microsoft. Oracle has remained competitive within the global software market, in part, because of its strategic acquisitions. One of the company's largest acquisitions took place when it acquired Sun Microsystems in 2010. While the company's net profits dipped in 2001 after the dot-com bubble burst, the company has enjoyed a steady rise in profits since that time, with a noticeable spike in profits between 2010 and 2013. As such, the company's profitability can be directly tied to its acquisition of Sun Microsystems.

Prior to its acquisition by Oracle in 2010, Sun Microsystems provided network computing infrastructure solutions, which included software, systems, storage, and microelectronics. In 2009, the final year of its independent operation, Sun reported approximately $11.45 billion in revenues and employed approximately 29,000 employees in more than 100 different countries. The lion's share of the company's revenues (42%) came from its Systems operating segment, which included the sale of servers that provide computing and storage power to customers as a key part of Internet infrastructure. The other core brands owned by Sun Microsystems were the Java technology platform, the Solaris Operating System, MySQL database management software, Sun StorageTek storage solutions and the UltraSPARC processor. Because the company relied on the provision of infrastructure-based services and products, the company was a large supporter of interoperability. Interoperability, here, is simply defined as the ability for different programs to exchange data with one another by using common formats. To facilitate innovation and interoperability, Sun made its key intellectual properties freely available as a strategy to support open standards, open interfaces, and open source software. By making a commitment to open source, Sun was viewed favourably by the open source community and maintained a relatively good relationship with the community because it was transparent about its corporate goals. To better understand the reasons for Sun open-sourcing some of their key intellectual properties, we need to consider some of the historical development for corporate involvement in FLOSS projects.

5.1.1. A Brief History of the Market for Operating Systems

Throughout the 1980s, the market for operating systems was dominated by proprietary versions of Unix-based operating systems. For example, Hewlett Packard offered HPUX, IBM offered AIX, and Sun Microsystems offered SunOS. These operating systems dominated high computing, or infrastructural level computing, while the consumer market was dominated by Microsoft DOS, which was not based on Unix but developed entirely by Microsoft. Importantly, the proprietary Unix-based systems were source-incompatible. In effect, although these systems were all based on Unix, the development of separate proprietary versions had caused the code to diverge in such a way that programmers could no longer assume interoperability between the systems. As a result, programmers had to maintain separate code bases for each system, and companies could sell entire stacks of software to their customers who had to accept the entire stack. This resulted in an inefficient system that was dominated by proprietary software vendors, while simultaneously increasing the workload for programmers. During the mid-1980s, however, the Free Software Foundation began as a response to the overly protective intellectual property restrictions placed on software. This, in turn, led to the development of free and open source software, which was collaboratively developed as a commons-based resource for others to study, use, adapt, or modify in any way.

Because this model of development was so successful, by the mid-1990s Linux, an open source operating system, had become the dominant Unix-like operating system. Linux undercut the competition by offering a comparable product at a significantly lower cost. Furthermore, because Linux is distributed under the GNU General Public License (GPL), an alternative form of intellectual property ('copyleft'), improvements to Linux could be shared by everyone, which improved its quality and stability. The proprietary companies could not compete with Linux because the commons-based peer production driving it constituted a larger labour force than any of the individual companies could employ. Rather than competing directly with Linux, certain proprietary companies began to open source their products as a way of joining forces with the free and open source software community. Sun Microsystems was one of those companies. Although Sun supported many different open source projects, I will focus on just three here. Sun open-sourced their Solaris operating system, which became OpenSolaris. They also open-sourced the MySQL database management software, as well as StarOffice, which became OpenOffice. As I mentioned earlier, Sun maintained a good relationship with the broader FLOSS community because of their commitment to and support for FLOSS projects. After the company was acquired by Oracle, this relationship was strained in certain ways. In what follows, I will discuss how the developers working on the three projects mentioned above – OpenSolaris, MySQL, and OpenOffice – strategically resisted the corporate acquisition.

5.1.2. OpenSolaris

In 1987, Sun Microsystems and AT&T announced that they were going to merge some of the most popular Unix-based operating systems into a single project. This project eventually became Solaris, which was a proprietary operating system held by Sun that contained both open-source and closed-source components. To attract interest in the project and build a community of users and developers around it, Sun Microsystems created OpenSolaris. OpenSolaris was an open-source version of the Solaris operating system, although it did contain some elements in its code that were not open source. After attracting a larger community of interest to the project, a Community Advisory Board (CAB) was created to direct it. The CAB served as a boundary organisation for negotiating boundaries between the OpenSolaris community and Sun. The

CAB was comprised of two Sun employees, two members who were elected by the broader community, and one member who was appointed by Sun from the broader free software community. In effect, most of the CAB members were connected with or appointed by Sun, and Sun made clear what its intentions were for the OpenSolaris project.

Sun's strategy for the OpenSolaris project was to incorporate some of the developments from OpenSolaris into their proprietary Solaris operating system. In turn, Sun could sell the proprietary version of Solaris to other enterprises. The money earned from sales of the Solaris project could then be used to support the developers and community involved in the OpenSolaris project. To facilitate this type of strategy, Sun protected OpenSolaris under a free software license created by the company called the Common Development and Distribution License (CDDL). This license enabled Sun to include proprietary, free software, or software protected under any other license in their Solaris and OpenSolaris operating systems. Consequently, Sun could use the OpenSolaris community as a way to drive development, quality control, or innovation that could be included in their proprietary Solaris offering. Importantly, however, Sun made this strategy very clear to the OpenSolaris community and was supportive of the broader FLOSS community, which gave it a good reputation within the community. Once they acquired Sun, Oracle took a very different approach to this strategy.

After Oracle acquired Sun, they announced plans to discontinue the regular distribution and development model of OpenSolaris (Laishram, 2010). Instead, Oracle would focus its development strategy on a new proprietary version of Solaris called Solaris Express. In effect, the new strategy from Oracle would not allow the community of developers that supported OpenSolaris to continue their work. In response, the Community Advisory Board directing the OpenSolaris project decided to fork the project. When a project is forked, developers take a copy of the source code and begin to develop it as a distinct form of software. The resulting fork of the OpenSolaris project is called OpenIndiana, which was created to continue the development and distribution of the OpenSolaris project. Currently, Oracle still continues development on the proprietary Solaris Express operating system, while the community of developers supporting OpenSolaris have left Oracle to work on the forked version of OpenSolaris called OpenIndiana.

In the case of the OpenSolaris operating system, Oracle's strategy was simply to discontinue the open source project and focus development on a proprietary version of Solaris under the new name Solaris Express. This represents the most direct strategy for ending open development. Oracle announced that the open source project would be discontinued and, in response, the community had to fork the project to continue development under a new name. This also illustrates how a FLOSS community can also continue working on a project even after production on a corporate-sponsored project was abandoned. This is a similar fate to that of MySQL and OpenOffice, but Oracle's strategy for ending development took different forms in each case.

5.1.3. MySQL

In 2008, Sun Microsystems acquired MySQL AB for approximately $1 billion (PC World, 2008). At the time, MySQL was growing in the market for relational database management software (RDBMS), and Sun's acquisition of MySQL would allow the company to compete directly with Oracle in that particular market. Only one year later, however, Oracle acquired Sun, and MySQL was one of the key properties that drew Oracle's interest. Indeed, the Sun-Oracle merger was originally approved by regulators in the United States, but the European Union (EU) did not immediately approve the deal specifically because of concerns that Oracle's acquisition of the MySQL property would lead to an anti-competitive market for RDBMS in Europe (Bloomberg, 2013). Consequently, the EU pressured Oracle to divest itself of the MySQL property as a condition for approval of the merger. As leaked documents provided to the whistleblowing site WikiLeaks have since shown, the United States Department of Justice communicated directly with the European Commission's Directorate General for Competition in support of the merger in October of 2009 (United States Mission to European Union, 2009). Less than three months later, in December of 2009, the merger was approved without the divestiture conditions sought by the EU.

MySQL relied on a dual licensing approach that was similar to the licensing of OpenSolaris. The dual licence model for MySQL would allow the code base for MySQL to be protected by the GNU GPL copyleft licence, but proprietary versions could be created for enterprises that wanted customised installations. When the Sun-Oracle merger was approved, employees working for MySQL had reservations about Oracle's intentions for the GPL-protected code base of MySQL. Most notable among them was Michael 'Monty' Widenius who authored the original version of MySQL and co-founded MySQL AB, which was the original owner of MySQL. Widenius later sold MySQL AB to Sun before Sun was acquired by Oracle. Widenius along with other MySQL developers were concerned that Oracle would try to discontinue MySQL or make it a closed-source program by using the same strategy it had with OpenSolaris. In response, Widenius urged MySQL users to 'Help MySQL' by starting an online petition. Leading up to the acquisition of Sun, however, Oracle pledged to keep the same licensing strategies in place that had been negotiated with current customers for an additional five years (Whitney, 2009). That commitment expired in December of 2014.

Fuelled by the concerns about Oracle's intentions for MySQL, the developers forked the project to create MariaDB.[35] The code base for MariaDB is protected by the GNU GPL, and is designed to be a drop-in replacement for MySQL.

As a forked project of MySQL, MariaDB allows its community of developers and users to ensure that the code will continue to be protected by the GNU GPL regardless of what Oracle decides to do with MySQL. Furthermore, although MySQL remains dominant in the RDBMS market with an approximately 58% market share, MariaDB grew to claim approximately 18% of the market (Fydorenchyk, 2014). MariaDB has experienced increased growth in the database market in part because of some notable companies switching from MySQL to MariaDB, including Google and the Wikimedia Foundation.

MariaDB once again illustrates how the community of developers and users of open source software can protect their projects from unwanted corporate encroachment. In the case of MariaDB, the project has gained additional attention from some of Oracle's competitors who have invested directly in it. Most notably, SkySQL recently invested nearly $20 million to support the growth of MariaDB. Backed by capital from Intel and from other venture capital firms, SkySQL is directed by some of the founding members of MySQL as well as former Sun executives who left the company after Oracle acquired the project. SkySQL announced a merger with The Monty Program AB, which is led by Monty Widenius, the original author of MySQL. The merger reunites the original members of MySQL and transfers ownership of the MariaDB trademark to SkySQL. The resulting partnership will focus on developing MariaDB to compete with MySQL.

Furthermore, both the Monty Program AB and SkySQL belong to the MariaDB Foundation. The MariaDB Foundation is a non-stock, non-profit corporation, which was established to provide legal and technical support for the MariaDB project and to provide a platform for supporters to contribute money to the project. For example, the MariaDB Foundation sells corporate memberships ranging from $5,000 to $100,000. According to the Foundation's web site, corporate memberships allow for the 'best opportunity to influence the future and present a point of view', although no further details are provided about exactly what that entails (MariaDB Foundation, 2018).

In sum, MariaDB represents another example of how FLOSS communities maintain the ability to protect their commons-based resource against unwanted corporate influence. In this case, however, Oracle's strategy was not to discontinue the open source project, per se. Rather, Oracle's acquisition of Sun allowed the company to gain a greater share of the RDBMS market, and Sun's ownership of MySQL was one of the primary properties that attracted Oracle to acquire Sun. Although development of MySQL still continues under Oracle, many of the community members resigned from Sun, and Oracle's commitment to maintain the same licensing agreements for MySQL expired at the end of 2014. To resist what could ultimately have been a similar fate to that of OpenSolaris, the MySQL community forked the project to develop MariaDB. In this case, Oracle seemed to violate the moral economy of the FLOSS community, but the community coped with that unwanted influence by forking the project to continue development under better conditions. Again, this represents a moment when the FLOSS community asserted a specific politics in protecting their working conditions; the community abandoned development on MySQL and moved to MariaDB. Furthermore, MariaDB has the additional benefit of having received investment capital from some of Oracle's competitors, which ensures the survival of the project for at least the foreseeable future. By establishing the MariaDB Foundation, the community has a legally recognisable organisation to provide technical and legal support for the project, while also collecting additional donations to the project. In the third and final example provided in this chapter, I focus on a series of office productivity software that eventually led to another forked project.

5.1.4. StarOffice, OpenOffice, LibreOffice

During the dot-com bubble in the mid- to late-1990s, Sun Microsystems experienced dramatic growth that allowed the company to make some key acquisitions. In 1999, Sun acquired the German company, StarDivision which developed StarOffice. StarOffice was designed as proprietary office software featuring word processing, spreadsheet, presentation, drawing, database, and formula programs. When Sun acquired StarDivision, the company continued to develop StarOffice as proprietary software. However, Sun forked the project and relicensed the software so that the source code could be made open source under a free and open source licence. Once again, Sun's strategy was to use the newly open-sourced software, known as OpenOffice, to develop new features and fix bugs in the software. Then, the changes made to OpenOffice could be integrated into StarOffice, which contained certain proprietary elements. OpenOffice could continue to remain free to consumers, while Sun would try to monetise StarOffice by selling the software and services to customers who wanted the additional features. The upshot for Sun was the maintenance and support for essentially two different versions of the same software: OpenOffice 1.0 was a forked version of StarOffice 6.0, and Sun maintained the legal rights to both properties, although they were protected by different licences.

The early versions of OpenOffice were protected by the Sun Industry Standards Source License (SISSL) and the GNU Lesser General Public License (GNU LGPL). Later versions were protected by an updated version of the LGPL after Sun discontinued the SISSL. The LGPL was chosen because it had less restrictive requirements for integrating free and open source software components into proprietary versions of the software. Although a full discussion of the distinctions between free and open source software licences is beyond the scope of this chapter, the basic differences between the GNU General Public License (GPL) and the GNU LGPL can be summarised quickly. The GPL requires that any modified or derivative software produced using GPL-protected software as its base must be redistributed under the same licensing requirements. This ensures that free software remains free software rather than being exploited by commercial companies. The LGPL is a more permissive licence that allows free software elements to be incorporated into proprietary software. The only restriction on using LGPL-protected software is that the end-user must have the ability to modify the source code. By protecting OpenOffice in this way, Sun could ensure that developments in OpenOffice could be used in their proprietary StarOffice.

Thus, the symbiotic relationship between StarOffice and OpenOffice continued under Sun because Sun was transparent about what its intentions were for the two properties. Importantly, however, OpenOffice was governed by a Community Council comprised primarily of members from the broader OpenOffice community but also including a Sun employee as well. The Community Council effectively served as a boundary organisation (O'Mahony and Bechky, 2008) between the community and the corporation. The Sun member on the Community Council was responsible for communicating Sun's intentions to the community. Once again, however, this relationship was strained when Oracle acquired Sun in 2010.

Since Oracle had discontinued the OpenSolaris operating system, members of the OpenOffice Community Council decided to create The Document Foundation and fork the OpenOffice project under the name LibreOffice until Oracle made its intentions clear for the OpenOffice project. Both The Document Foundation and LibreOffice were established with the intention of being temporary projects until Oracle made its intentions clear. In the event that Oracle ultimately decided to discontinue OpenOffice, however, the Community Council would be able to move development to the newly created LibreOffice. Furthermore, The Document Foundation was established as a non-profit organisation to manage the LibreOffice project and promote the use of open source document software more broadly. The initial governance of The Document Foundation was directed by a temporary steering council featuring some of the same members of the OpenOffice Community Council. Oracle viewed the Community Council members' positions on two governing boards as a conflict of interest and asked members on the Community Council to step down from their positions (OpenOffice Community Council, 2010). This move effectively ended community support for OpenOffice and the project was renamed Oracle OpenOffice. Oracle OpenOffice became the proprietary software offering from Oracle that was meant to replace Sun's StarOffice.

While the official position of Oracle was to cite a conflict of interest, members of the broader open source community viewed Oracle's broader strategy as simply wanting to discontinue open source projects that existed under Sun because they did not provide any real value to the company. In effect, not only did the governance structure change under Oracle's ownership, but Oracle also seemed to have violated the moral economy (Thompson, 1971) of the FLOSS community. In response to this, however, The Document Foundation continued its development of LibreOffice. Since LibreOffice had strong community support, LibreOffice essentially surpassed OpenOffice within one release. In

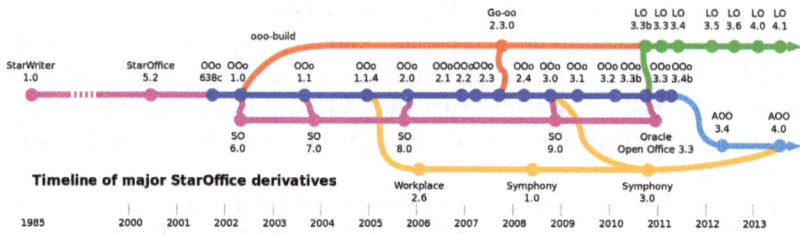

Figure 5.1: Major StarOffice Derivatives (image has been released to the public domain and is available from https://en.wikipedia.org/wiki/StarOffice#/media/File:StarOffice_major_derivatives.svg)

effect, all of the collective labour behind the development of OpenOffice abandoned the project but continued to work on LibreOffice. Because OpenOffice had been abandoned, Oracle announced that it would end development on the project entirely and fire the majority of OpenOffice developers. Ultimately, Oracle donated the code base for OpenOffice to The Apache Software Foundation, which has resumed development on the project under the name Apache OpenOffice.

To summarise this somewhat confusing history of a software that has been forked numerous times, Figure 5.1 illustrates the development history of StarOffice, its transition to OpenOffice (OOo) under Sun, the dual development of StarOffice (SO) alongside OpenOffice, the forks into LibreOffice (LO) and Oracle OpenOffice after Oracle acquired Sun in 2010, and the donation of OpenOffice back to The Apache Software Foundation to be developed as Apache OpenOffice (AOO). Figure 5.1 also includes additional forked projects that have not been discussed in this chapter, which include IBM Lotus Symphony (Symphony) and Go Open Office (Go-oo). As illustrated in the figure, the developments offer examples of how the FLOSS community uses legal, technical, and governance strategies to protect their commons-based resources.

5.2. Protecting the Commons

Throughout this chapter, I have demonstrated how the FLOSS community maintains the ability to leverage its collective labour power against undue corporate influence by employing technical, legal, and governance strategies to protect its commons-based resources. On the one hand, FLOSS has unique technical characteristics that allow it to be reproduced and distributed widely without any significant cost. This allows FLOSS projects to be forked so that development can occur collaboratively, simultaneously, and continuously throughout the life of the project. Although dispersed development occurs, however, the community employs certain governance strategies for effectively coordinating development and protection of the project. These governance strategies include the establishment of non-profit organisations, which hold the intellectual properties for projects. These organisations provide a legally recognisable entity that can more effectively defend the intellectual property and licensing requirements of the project. Furthermore, more direct governance of the development project can occur through governing councils that are democratically elected or appointed by the community.

The legal strategies for defending FLOSS projects rely on alternative intellectual property protections like copyleft or other free and open source software licences. These licences free the software from overly protective copyright and allow the community to fork the project in the event of undue corporate influence. On the other hand, corporations can also use licensing strategies to their benefit as well. In the case of Sun, the company used licensing that allowed for free and open source software development but that was less restrictive to the corporation. These licences allowed the company to incorporate some of the commons-based peer production of FLOSS projects into their proprietary offerings. This strategy was understood and accepted by the FLOSS community because Sun was clear about its strategies but also because Sun supported FLOSS development projects. In a sense, then, licensing a project becomes a site of struggle, especially because a single project may contain code that is protected by different licences. These licences may have competing or conflicting terms that need to be resolved or the project becomes susceptible to intellectual property litigation. As was the case during Oracle's acquisition of Sun, the licences can be changed as a way to direct development toward different ends. Sun was transparent about its licensing strategies as a part of its broader commercial strategies, while Oracle made either temporary commitments to use existing licensing strategies (e.g. MySQL) or sought to change those licensing requirements altogether (e.g. OpenSolaris).

However, the dynamics that exist between FLOSS communities and corporations are comprised of a combination of technical, legal, and governance strategies. The particular forms that these strategies take will vary depending on the individual project, but the FLOSS community's ability to defend its commons-based resources depends, in part, on a shared consciousness of what is permissible within the community. In a sense, this shared consciousness constitutes a sort of moral economy (Thompson, 1971). The FLOSS community leverages its collective labour power against corporate power by protecting its commons-based resources. When a corporation infringes on the moral economy of the community, the community rebels by forking the project and abandoning the project that has been overly influenced by the corporation. This moral economy has foundations in the shared ideals of peer-to-peer relationship building, collaborative development, transparency, and community.

Even though the FLOSS community maintains the ability to leverage its power against undue corporate influence, community members are still in a somewhat precarious position as digital labourers. One definition of success in open source projects is to receive backing from a company, which at least ensures the project's survival if not its overall attractiveness. However, the FLOSS community depends on keeping projects protected under free software licences, albeit of many different types, so that the community maintains the ability to keep the code for the program open. This is particularly true in cases where hybrid models of proprietary and free software are used in FLOSS projects. Throughout this paper, I have demonstrated how such struggles can occur, particularly after corporate mergers, acquisitions, or takeovers.

In the face of growing corporate involvement in FLOSS projects, the broader FLOSS community must maintain its ability to protect its commons-based resources. At the same time, however, the protection of these resources depends, at least in part, on a shared collective understanding of how the community can leverage its collective labour power against increasing corporate involvement. The lessons to be learned from Oracle's acquisition of Sun Microsystems need to remain salient if similar strategies are to be effective. Most important, however, is the recognition that the struggles taking place within the FLOSS community are just one part of a broader social struggle. As Christian Fuchs (2008) has observed, commons-based production is not truly possible until we have a commons-based society. Until that time, commons-based movements like FLOSS will be subjected to increasing corporate encroachment that threatens to abate, assimilate, or altogether annihilate progress toward alternative economic configurations.

Notes

[33] An earlier version of this chapter appeared as Benjamin Birkinbine 2016b. Conflict in the Commons: Toward a Political Economy of Corporate Involvement in Free and Open Source Software. *The Political Economy of Communication* 2(2): 3–19.

[34] Unless otherwise noted, all of this information was derived from Oracle's annual filings (Form 10-K) with the Securities and Exchange Commission (SEC) of the United States, which is available here: https://investor.oracle.com/financial-reporting/sec-filings/default.aspx (last accessed 2 January 2019)

[35] MariaDB is just one fork of the MySQL project. Percona Server is another that is still actively developed as of the time of writing.

CHAPTER 6

Conclusion: From Capital to Commoning

'...the new technology is itself a product of a particular social system, and will be developed as an apparently autonomous process of innovation only to the extent that we fail to identify and challenge its real agencies. But it is not only a question of identity and defence. There are contradictory factors, in the whole social development, that may make it possible to use some or all of the new technology for purposes quite different from those of the existing social order: certainly locally and perhaps more generally. The choices and uses actually made will in any case be part of a more general process of social development, social growth and social struggle.'

(Williams, 1975: 135–136)

The quote from Raymond Williams above emphasises the contradictions inherent in the ways in which new technologies are put to use. On the one hand, new technological developments may usher in a period of optimism or utopian thinking when assessing the potential uses of the technology. On the other hand, new technologies are also susceptible to co-optation by existing power structures. In this sense, all technology is dialectically situated within 'a general process of social development, social growth, and social struggle.' The goal of this struggle, especially for those interested in finding alternatives to the prevailing system, is to find ways of changing existing power structures to advance the cause of human dignity, mutual aid, trust, and conviviality.

The purpose of this book was to demonstrate how one such technology – free and open source software – is dialectically situated between the commons and capital. To illuminate the ways in which these forces struggle over free and open source software, my task was to 'identify and challenge' the 'real agencies' of free and open source software as commons under capitalism. In doing so, I identified the specific ways in which capital incorporated the forces of commons-based peer production into capitalist enterprises, the motivations for doing so, and the ways in which communities of free and open source

How to cite this book chapter:
Birkinbine B. J. 2020. *Incorporating the Digital Commons: Corporate Involvement in Free and Open Source Software.* Pp. 101–119. London: University of Westminster Press. DOI: https://doi.org/10.16997/book39.f. License: CC-BY-NC-ND 4.0

software developers cope with unwanted interference in their projects. Moreover, I approached this study historically, paying close attention to the historical forces that enabled both the rise of commons-based peer production as well as the incorporation of those forces into capitalist production. In this concluding chapter, I summarise some of the main findings from the case studies and reflect on their significance for advancing the commons under capitalism.

6.1. Major Findings

This study complicates and extends theorisations of commons-based peer production by investigating sites where the idealism of FLOSS production meets with the material realities of capitalism. These contested sites make up the case studies in this research project, for they are where commons-based peer production has been incorporated into the corporate structures of capitalist firms. By employing a critical political economic approach, this study focused on the power relations that exist between corporations that rely on capitalist, market-driven production, and the broader FLOSS communities that rely on non-market, commons-based peer production. An important part of this focus was to position the commons and capitalism as operating according to different systems of value. At times, these two systems are capable of working together by coupling through the commodity form. The processes of commodification were demonstrated in those case studies that illustrated how FLOSS projects have been incorporated into commercial offerings. However, at other times, these systems diverge, which can lead to an antagonistic relationship between capital and the commons.

In previous literature, major projects like the Linux kernel or Wikipedia have been lauded as examples of effective and productive commons-based peer production (Benkler, 2006; Lessig, 2006; Weber, 2004). Significantly less studied, however, is how capitalist firms can use commons-based peer production to supplement their commercial offerings. The case studies for this project, particularly the discussion of Red Hat and Sun Microsystems, provided an in-depth look at how capitalist firms rely on the innovations and bug fixes from within the FLOSS community for implementation in their commercial products. That said, however, these case studies should not necessarily be viewed as generalisable across all FLOSS projects. The broader ecosystem of FLOSS projects features certain projects that are completely supported by their community of developers and do not rely on investment or sponsorship from corporate firms.

By selecting cases in which capitalist firms are incorporating commons-based peer production, this study was able to yield a novel insight into how intellectual property is used both within the FLOSS community and corporations. Specifically, the case of Red Hat demonstrated how a firm is able to profit from intellectual property that is covered by the GPL and, therefore, not amenable to enclosure by traditional copyright. Because Red Hat cannot exclude others from using its source code by relying on copyright, the company uses its trademarks to prohibit competitors from making a direct use of its products. However, Red Hat's trademarks cannot prevent someone from using the underlying source code, which is protected by copyleft. Indeed, this was the case with CentOS, which was designed as a functionally equivalent operating system to that offered by Red Hat Enterprise Linux, Red Hat's core commercial product. Similarly, Red Hat controls the types of licences that can be included in its Fedora Project, which is the FLOSS project that generates the code included in its commercial offerings. The ways in which Red Hat controls the intellectual property included in its commercial offerings complicates the claims made about the productive autonomy within FLOSS communities.

In the vast majority of work on FLOSS, one of the defining features of its novelty is often traced back to its protection under more permissive copyright licences, or copyleft licences (Lessig, 2001; Stallman, 2002; Benkler, 2006). In addition, the software industry has been broadly plagued by a surge in patent infringement claims. However, the issue of trademark is an often-overlooked feature of software development. Red Hat uses trademark protections to circumvent the permissive nature of the GPL and the other licences that do not allow it to claim exclusive ownership of the code used in its core products. Although Red Hat is just one example and, perhaps, an exceptional one, the case serves as a contradictory example to the overarching claims made about the degrees of freedom, democracy, and autonomy within FLOSS production.

Further complicating these claims are the often-overlooked Contributor Licensing Agreements within FLOSS production, particularly when a project has a corporate or other institutional sponsor. While these agreements are not uniform across all FLOSS projects, the organisations that issue them rely on these agreements to maintain control over their projects. However, control is achieved in at least a couple of different ways. The CLAs may ask contributors to surrender the rights to their submissions so that the organisation can defend itself from intellectual property claims. Similarly, the CLAs may be used to control the types of licences that are allowed into the code base. This was seen in the Red Hat case study, whereby Red Hat wanted to guarantee to its customers that they would not be in danger of intellectual property infringement suits. A common theme running throughout the Red Hat chapter was the extent to which copyright separates authorship from ownership. In this sense, the current project contributes to this critical understanding of copyright by demonstrating how FLOSS labourers are forced to abandon claims to ownership of their work in order to contribute directly to certain FLOSS projects.

6.2. Case Studies

Each of the case studies presented here provides lessons for understanding the relationship between capitalism and the commons. The cases chosen were purposely selected because of their prominence within both corporate and FLOSS communities. Red Hat, Microsoft, and Oracle represent some of the largest and most publicly visible software companies in the world. This is primarily the reason for selecting these companies, but also means that the findings from each case study may not be applicable to a broader range of corporations or FLOSS projects. Furthermore, not all FLOSS projects have corporate sponsors. In this sense, the study provides a snapshot of the ways in which corporations incorporate the FLOSS commons. When considered together, however, these case studies illuminate some of the general dynamics occurring at the intersection of corporations and the commons. In what follows, I discuss the more specific implications of each case study for understanding this phenomenon.

6.2.1. Microsoft Corporation

Microsoft has a long history of opposition to FLOSS. This stance began as early as 1976 when Bill Gates authored the 'Open Letter to Hobbyists', in which he railed against the culture of sharing software within the community. He argued that this practice harmed the ability of others to produce software and be compensated for their work. However, this stance contradicts some of Microsoft's own history, as it relied on others' designs to produce some of its most successful software. This was particularly the case for the MS-DOS operating system and the graphical user interface of Windows, which were built on top of previously existing technologies developed in Gary Kildall's CP/M operating system and Apple's graphical user interface. Both of these technologies were instrumental to Microsoft's success throughout the 1980s and 1990s, especially when paired with its strategic partnerships with IBM and other OEMs, which allowed the company to gain widespread adoption of its software. The same can be said of its Internet Explorer web browser, which the company packaged with distribution of its Windows operating system. This practice ensured that the company's web browser would win the first of the Browser Wars, but it also was one of the primary business practices that led to its conviction for antitrust violations by the Department of Justice.

Microsoft's ascent to the top of the personal computer software market culminated around the same time that it was being investigated for antitrust violations. When the United States Department of Justice (DOJ) issued its decree in 2001, Microsoft was forced to divest its operating system and applications operations. However, after the original District Court judge recused himself from the case after making some public comments that gave the impression of bias against Microsoft, the subsequent judge no longer sought divestment. Rather, Microsoft needed to agree to a series of consent decrees that were designed to prevent the type of predatory and non-competitive behaviours that led to its conviction. The consent decrees were intended to last for five years, but they were renewed twice and finally came to an end in 2011. However, the decrees did little to affect Microsoft's economic performance, as the company's annual revenues and profits continued to climb in the wake of the DOJ's decision. Nevertheless, as argued in Chapter 3, the antitrust suit marks a major historical moment both for Microsoft and the software industry more generally. Most notably, the antitrust suit forced Microsoft to make its APIs more openly available to other developers so they could design software that could interact with Microsoft's technologies. The antitrust decision also coincided with the bursting of the dot-com bubble in 2001, the emergence of Linux as a commercially viable business model, and the emergence of the so-called Web 2.0 era, which shifted the business focus of many high-tech companies during that period.

The antitrust conviction also signalled to Microsoft that it needed to find new ways of doing business. Because Linux was becoming more widespread, Microsoft could no longer take an antagonistic stance toward open source. Instead, it needed to find ways to ensure that its products could function on devices that use Linux. To facilitate greater interoperability between Microsoft and non-Microsoft technologies, Microsoft expanded its Shared Source program and, in 2012, opened an entire division of the company dedicated to promoting and supporting open source, open standards, and open platforms. The trend toward embracing open source software continued even after Microsoft closed its Microsoft Open Technologies division, as the company now claims that it is unnecessary to have a separate division devoted to open source. Rather, they argue that open source has become instrumental to everything they do. Indeed, Microsoft also purchased GitHub, the world's leading software development platform, which is used primarily to host open source software projects. This shift in Microsoft's stance toward open source is indicative of the fact that FLOSS, by many measures, has proven to be an effective and commercially viable production model. The shift in supporting open source projects suggests that Microsoft is trying to accomplish two primary goals: harnessing the power of commons-based peer production to supplement its own commercial goals as well as promoting interoperability between its technologies and other systems.

The Microsoft case study is indicative of a company undergoing a transformation in its stance toward FLOSS. In part, this shift was driven by the antitrust conviction in 2001, but the leaked Halloween Documents suggest that the company was already concerned with the FLOSS phenomenon and how to combat it in 1998. Perhaps not coincidentally, this is the same year that the antitrust investigation began. The Microsoft case study is useful for understanding the relationship between FLOSS and corporations because of Microsoft's dominance of the software market. As such, it is instructive to trace its history of software development, especially since the company spans both the 'Web 1.0' and 'Web 2.0' eras. During this time, its business practices and overall strategy shifted to take advantage of the emerging threat of FLOSS development. The company sought ways to incorporate the commons into its existing business operations in part because of the antitrust convictions but also because FLOSS development was proving to be a successful competitor to the company's own development practices.

6.2.2. Red Hat, Inc.

In the case of Red Hat, which still maintains a relatively good relationship with the FLOSS community, the company was able to harness (which is to say, centralise) the collective labour power of the FLOSS community and transform it into a profitable business strategy. Red Hat was created with the intention of providing a formalised institution that could bring the power of free software to the market. However, since the underlying source code for free software was protected by the GNU General Public License (GPL), Red Hat was unable to rely on using copyright protection to exclude others from providing similar software and services. As a result, the company began offering customised versions of free software that could be packaged and protected under the Red Hat corporate logo. As such, the company's products could be protected by trademark. The software that the company provides, then, is protected by the Red Hat trademark, and the company sells customised subscriptions for its software and services. However, Red Hat still needed a way to protect its customers against potential intellectual property infringement claims. Consequently, the company needed a way to control the types of licences allowed in its software offerings. To accomplish this, Red Hat first required all contributors to its software to sign an Individual Contributor License Agreement (ICLA), which would assign the rights to protect the code to the company. The ICLA later changed to the Fedora Project Contributor Agreement (FPCA), which served as a mechanism to control the range of possible licences that could be included in contributions to its Fedora project. Nonetheless, the consequence of controlling the commons was the same.

From one point of view, Red Hat might be viewed as a pragmatic solution to the problem of organising commons-based peer production so that it can become conducive to the establishment of a capitalist enterprise. In effect, Red Hat serves as a formal organisation that can accept liability for the products and services it provides to other businesses. In other words, the problem of organising commons-based peer production under capitalism was solved by establishing a legally recognisable and formal institution that serves as a mediator between corporations and the commons. To accomplish this, however, Red Hat needed to find a way to control what types of code – or at least the types of intellectual property licences – were included in its software so that it could protect itself and its clients against intellectual property infringement claims.

In this sense, Red Hat functions as a curator of the commons. Just as a curator is responsible for collecting, organising, and interpreting artefacts for the purpose of public display, Red Hat performs a similar function for its subscribers. In each case, the curator charges a fee to the public for entrance to a purposefully organised and constructed display of artefacts that has been interpreted in a particular way. The key difference, however, is that Red Hat does not rely on the collection of artefacts exactly as they existed previously. Rather, Red Hat relies on commons-based peer production from its FLOSS project, Fedora, for inclusion into its customised distributions of Red Hat Enterprise Linux. Moreover, the contributions to Fedora are controlled by worker agreements that all contributors to the Fedora Project must sign. Importantly, however, because Red Hat is transparent about its intentions, the company has been able to enjoy a relatively good relationship with the broader FLOSS community throughout its history.

Whereas Red Hat is situated as a mediator between corporations and the commons of free software production, the Fedora Project Board also serves as a boundary organisation (O'Mahony and Bechky, 2008) between the community of programmers who contribute to the Fedora Project and Red Hat. As such, it is here where the boundaries between Red Hat and the Fedora Project community are negotiated. Similar organisations exist in other FLOSS projects and serve as a useful mechanism for negotiating the boundaries between capital and the commons. Through these processes, as well as the mechanisms used by Red Hat to use FLOSS production as part of its business model, the Red Hat case study represents the ways in which the value of FLOSS production can move from the commons to capital.

6.2.3. Oracle's Acquisition of Sun Microsystems

The third case study, Oracle's acquisition of Sun Microsystems, most directly addresses the question of what happens when a corporation exerts unwanted influence on a FLOSS project as well as how a FLOSS community can cope with the unwanted influence. The chapter illustrated how the FLOSS community has coped with undue corporate influence into its projects by focusing on three different FLOSS projects that were supported by Sun Microsystems prior to its acquisition by Oracle: the OpenSolaris operating system, the MySQL relational database management system, and the OpenOffice office productivity suite of software. What becomes clear from the case study is that FLOSS projects may not be able to avoid corporate influence altogether, especially when those projects are sponsored or supported by a particular company. However, given the nature of FLOSS code, the FLOSS community maintains the ability to abandon production on a particular FLOSS project by forking the project and continuing development under a new name. This is precisely what happened in each of the three cases discussed in Chapter 5.

Furthermore, the case study also provides evidence that FLOSS projects are not immune from the corporate manoeuvering – acquisitions, integration, takeovers, buyouts etc. – that is commonplace in a capitalist system. That is, although the projects may find a corporation willing to provide support through sponsorship, financing, or partnerships, those relations can become strained in the wake of an acquisition in which the acquiring company is unwilling to provide the same level of support as the previous company. If this is the case, the community of developers who contribute to the FLOSS project have technical, legal, and governance strategies at their disposal to resist undue corporate influence in the project. Technically, code can be reproduced ad infinitum without any substantial reinvestment costs. Legally, most code that is used in FLOSS projects is protected by permissive licences that allow the community to fork their project and begin development under a new name. Coinciding with the process of forking the project is the transitioning of the governing board members to oversee the new project.

The Oracle Corporation's acquisition of Sun Microsystems illustrates how the power dynamics existing between FLOSS communities and the corporations that rely on their projects are complex and varied. While the community still retains the power to abandon production on a project in the face of undue corporate influence, this still places the community in a precarious position with respect to the long-term survivability of their projects. The community retains the ability to fork the project and begin new development, but it cannot rely on the same level of support it received from its corporate sponsor unless it can find new investors. For instance, the OpenIndiana, MariaDB, and LibreOffice projects were able to find additional investment capital, although to varying degrees. In other words, the ability to fork a project is just one step in assuring productive autonomy. However, the productive autonomy of those who contribute to projects that are sponsored by other organisations may always be at risk of undue influence. In those situations, the community can take steps to try to reduce such influence.

6.3. On the Benefit of the Commons Paradigm

In extrapolating from the lessons learned in these three case studies, we can also draw some lessons for the commons more generally, especially the commons paradigm that has been used to understand FLOSS production and reproduction. The benefits of the commons paradigm can be summarised in three different ways that are all interconnected. The commons paradigm is simultaneously universal, adaptable, and teleological.

It is *universal* in the sense that it establishes a framework for understanding how collective resources ought to be governed to ensure their survival and reproduction over time. This framework can be used by any commons-based movement regardless of the unique conditions within any local context. Indeed, various commons movements can learn from what other commons-based movements are doing, then make a decision as to whether such a change should be implemented within their own governance structures.

In this sense, the commons framework is also *adaptable* in that it provides a flexible framework that can be applied across a variety of social struggles. In other words, it is not normative in that it does not posit only one way of accomplishing collective governance. This is perhaps why it has found currency within autonomist Marxism. The autonomous approach focuses on workers' ability to define themselves independently from capital, while focusing on the different strategies for resistance that are possible in all aspects of social life. The specific dynamics of each community's struggle, however, are determined by a couple of different factors. First, these struggles are confronted with local, national, regional, international, or global forces that shape the institutional or political–economic arrangements within which each community is situated. These forces are not mutually exclusive categories; rather, the struggles may be shaped by some combination of these broader forces. Second, these struggles are also shaped by the unique historical, social, and cultural dynamics of each community. Within FLOSS production, the primary concern is creative autonomy, but other communities connect the survival of commons-based resources with the survival of an entire way of living within ecological contexts. Regardless of the particular struggle, commons-based movements generally want to preserve their shared resources from exploitation or destruction.

Finally, the commons paradigm is *teleological* in that it helps us imagine a post-capitalist future that is on the horizon. As was discussed in Chapter 2, commons movements and the activity of commoning can be understood as 'ways of becoming,' denoting a process by which social change is possible. As such, they serve the purpose of demonstrating the ways in which an alternative future is possible. Commons movements rely on the shared values of mutual aid, trust, conviviality, cooperation, and solidarity. Moreover, these values are also intertwined with the complex histories, cultures, and ecologies of the communities within which they are situated. These values are antithetical to capitalism, which values profit maximisation, self-interest, and competition. The question remains, however, as to how we can continue to build commons-based movements, as well as linking them together so their collective power no longer remains fragmented.

6.4. Political Organisation from Below

There is a contradiction that exists today for organising political resistance.[36] On the one hand, the spread of digital technologies has assisted diverse and fragmented publics in linking with others to form networked communities of interest. Such communities, like those involved in free software projects, rely on inputs from a distributed community of contributors who can collaboratively produce goods, services, or create new meanings for cultural texts. On the other hand, these communities continue to operate from within existing institutions, which operate according to liberal-democratic logics. These networked publics have challenged previously held assumptions.[37] As just two examples of this, consider the challenge to assumptions about ownership (i.e. the rise of copyleft licences to challenge traditional copyright protection), and to production bounded to a specific nation-state and its regulatory policies (i.e. globalised commodity supply chains and the question of whether a product is 'Made in the USA' or any other single country).

This raises the question of what organisational form political resistance should take from within this context. On the one hand, we want to preserve the relative autonomy of local communities to organise in ways that make the most sense for the community. On the other hand, we are confronted with existing institutions that require the coordination of diverse movements to effect change within those institutions. As it concerns the digital commons, Dulong de Rosnay and Musiani (2016) have developed a typology of centralised versus decentralised peer production that is instructive here. The typology can be seen below in Table 6.1. The goal for the digital commons would be to move increasingly toward the decentralised models presented in the table. Doing so would allow local communities to respond to unique needs and simultaneously preserve the highest degree of autonomy for the community.

DuLong de Rosnay and Musiani (2016) are not the only scholars wrestling with how to advance decentralised peer production forward to mount a challenge to capitalism. One such debate took place in a series of articles published in *tripleC: Communication, Capitalism, and Critique* in 2014. The debate stemmed from a proposal made by Bauwens and Kostakis (2014). Noting the contradictions of commons-based peer production being co-opted by capitalist firms, as well as the growing co-operative movement and worker-owned enterprises, Bauwens and Kostakis (2014) propose a convergence that they call 'open co-operativism' that would 'combine Commons-oriented open peer production models with common ownership and governance models such as those of the co-operatives and the solidarity economic models' (356). To facilitate such a movement, the authors suggest the creation of an alternative intellectual property licence that would require reciprocity to benefit the commons. They frame this as a shift from a 'communist' licence like the GNU General Public License (GPL), which allows anyone – including capitalist firms – to use the commons-based resource, toward a 'socialist' commons-based reciprocal licence which, they argue, is exemplified by the Peer Production License (PPL) as proposed by Kleiner (2010). Such a licence would allow for commercial use

Table 6.1: Centralised Versus Decentralised Peer Production (Dulong de Rosnay and Musiani, 2016: 196).

	Ownership	Technology	Governance	Rights	Value
Centralised	Company Major platforms	Central server controlled by platform owner	Top-down decision-making by platform owner	Exclusive rights assigned to platform owner	Concentrated in platform owner
Decentralised	Cooperative non-profit Informal, unstructured collaboration	Several user-controlled computers/ nodes linked in a peer-to-peer network	Participative democracy Autonomy of peers	Terms of contribution leaving some rights to contributors	Redistributed within community and/or society at large

of the licenced resource, but would require reciprocity to the community. This means that licensing fees would be charged to for-profit companies that use the resource. This, then, would allow the community to establish a co-operative, which could receive the licensing fees as income that could then be used to maintain the commons. In effect, the goal is for the community to retain the surplus value of their production. The authors further argue that the goal of this project is to transform the mode of production toward the commons. Furthermore, they claim that without such a transition commons-based peer production 'would remain a parasitic modality dependent on the self-reproduction through capital' (Bauwens and Kostakis, 2014: 360).

Meretz (2014) critiqued Bauwens and Kostakis's proposal on a couple of fronts. First, Meretz critiques the 'logic of exclusion' embedded within the proposal for licensing. He argues that free software is not a commodity; it can be appropriated and used by everyone, but the GPL prevents its transformation into a commodity. Second, he critiques the authors' use of 'reciprocity' by claiming that licences are never reciprocal. Rather, licences grant or deny access or use. Reciprocity must involve people who are reciprocal in a social relationship.

Meretz's own view is that social transformation is not possible by building a counter-economy for progressive social movements. In his words, 'it is not possible to out-compete capitalism ... to be better than capitalism on its own terrain in order to finally get rid of it' (Meretz, 2014: 364). Rather, we need a *new social logic of producing our livelihood*, which will not be built upon existing logics of exclusion that mark commodity production. Indeed, capitalism must constantly open up spaces for new logics to emerge so that they can be exploited. In the end, Meretz views the proposal for a new socialist licence as a mechanism for *accessing* the economy rather than a means for societal transformation.

Rigi (2014) offers his own views on these proposals by revisiting some foundational concepts from Marx's work (i.e. value, profit, surplus value, and rent), then demonstrating how Bauwens and Kostakis fall short in their application of these concepts. His point is not to impose Marx's own views on Bauwens and Kostakis, but rather to suggest that they offer concrete definitions for how they use these terms, which would aid in the development of a theory. In addition, Rigi agrees with Meretz's claim that further engagement in the market economy on behalf of peer production communities would only lead to those practices being assimilated into capitalism. However, he also critiques Meretz for underestimating the communist nature of the GPL. Rigi's point is that the GPL already requires reciprocity by stipulating that any derivative work produced with GPL-licenced code must also be made available under the same licence. In this regard, Rigi argues that the GPL abolishes knowledge rent, as there is no 'owner' of the commons who can charge rent for using the commons. Furthermore, Rigi points to companies like IBM who decided to release their proprietary code to the commons so that it could be integrated with Linux. In so doing, the scope of available commons-based code expanded through the specific mechanism of the GPL.

In the final section of his article, Rigi (2014) outlines his own vision for how radical social transformation is possible. His goal is to examine how it would be possible to use the principles and lessons from the production of digital commons to revolutionise material production. Rigi identifies two fundamental problems that must be overcome for this to be possible: territorialisation and automation. First, the production of Linux can occur regardless of geographic location, and contributions to the digital commons can be shared easily across space in very little time. This is because anyone with access to a computer (and the necessary coding skills) is able to contribute to Linux or another FLOSS project. The same cannot be said of material production. Noting both the transportation and ecological costs associated with moving material production across space, Rigi concludes that any attempt at applying commons-based peer production to material production must be geographically bounded so that the production site is in close proximity to the consumption site. Second, material production is increasingly automated, and the human contribution in this sphere is increasingly relegated to science, design, and software. Therefore, 'a combination of a Linux mode of cooperation with automation will generalise peer production to all branches of production' (Rigi, 2014: 400). However, certain spheres of social life will remain untouched by automation: symbolic activities (like artistic expression, knowledge, etc.), and care for humans and nature (education, ensuring ecological survival, etc.). Rigi concludes his article with some speculative proposals for how we might bring about some of these changes by specifically arguing for something he calls 'revolutionary peer producing cooperatives'. I will revisit this proposal later in the conclusion, as it dovetails nicely with some of my own proposals.

In the meantime, however, one can begin to imagine how a set of diverse and distributed communities could begin to implement practices associated with commons-based peer production. Indeed, we have already seen examples of this around the world, but these communities still need to be linked through common interests to mount a significant challenge to existing institutions. This is where De Angelis's (2017) use of 'boundary commoning' becomes useful. In this final section, I outline how a commons praxis might overcome these two difficulties. First, I discuss the problem of organisational form by building upon lessons from recent critical scholarship. Second, I discuss 'subversive commoning', which would address the need for a progressive political project for moving the commons forward.

6.5. Boundary Commoning

De Angelis's (2017) formulation of circuits of commons value, which was discussed in detail in chapter 2, provides a useful analytical tool for understanding how value is produced and reproduced by commons-based movements. However, these movements still intersect with capital accumulation circuits in the course of their commoning activities. Therefore, the coupling of commons circuits of value with capital accumulation circuits, whether willingly or out of necessity, still does not overcome many of the contradictions of the commons. De Angelis's formulation, then, seems to leave us with a picture of a 'long social revolution', which would proceed primarily through the autonomous development of an emergent alternative value system from within capitalism. Such a value system would privilege commons value rather than capital accumulation. But there is another element in De Angelis's work that he draws from systems theory and cellular biology, which seems to contain the possibility of linking diverse commons movements. That is the concept of 'boundary commoning', which is defined as:

> the commoning that exists at the boundaries of the commons systems and that creates social forms of any scale, opens up the boundaries, establishes connections, and sustains commons ecologies, or that could reshape existing institutions from the ground up through commonalisation and create new ones. (De Angelis 2017: 24)

Boundary commoning has the potential to provide an organisational model for how diverse and distributed commons-based movements can work together toward a common goal. Through the multiplication of commoning activity and the interweaving of commons-based communities through boundary commoning, a commons movement could potentially lead to a tipping point at which social transformation is possible. In addition, De Angelis claims that commons movements could link with social movements to form a hybrid movement with the combined power to bring about social revolution. As he explains, these 'are not movements of fragmented subjectivities sharing a particular passion, but movements of connected subjectivities whose connection is further increased by their social movement' (Ibid., 387). Therefore, boundary commoning allows specific communities to retain their autonomy, while also linking with other organisations through common interests. While similar organisational structures have been used in the past – namely, the federated approach taken by Indymedia (see Pickard, 2006) – the commons offer a framework that is widely applicable and capable of linking diverse movements under a common framework. Importantly, however, such a movement ought to be based on an antagonistic understanding of the commons' relation to capitalism. In short, we continue to need a form of commons praxis for advancing the cause of the commons.

6.6. Commons Praxis

The task for a commons-based praxis is to overcome at least two hurdles. First is the task of determining an organisational form that would incorporate the lessons of critical scholarship on the commons. Critical scholarship has exposed some of the limitations of liberal-democratic or reformist approaches that seek to transition to a commons-based society from within existing institutions. While this is undoubtedly necessary to bring about change, we are still left with the limitation of radically transforming the organisation of society and social relations from within existing institutions, which are based on hierarchical organisational structures that tend to privilege political and economic elites with the requisite capital necessary to exercise influence by shaping policy agendas. These institutions cannot account for the multitude of distributed, diverse, and unique needs of local communities, and yet their existence will continue unless commons-based movements provide alternatives. This problem has become even more acute now that local publics can network with other communities of interest across national and international geographic boundaries. Second, a commons praxis needs to overcome the persistent problem of growing and sustaining commons-based movements over time. In this sense, a commons praxis needs to move beyond a *politics of subsistence* and institute a more progressive politics that would actively seek to grow the commonwealth available to commoners. I refer to this political project as 'subversive commoning'.

6.6.1. Subversive Commoning

The unique characteristics of the digital commons – low rivalry and low excludability – make it possible for the products of peer production to be appropriated by the state and capital. Similar arguments have been made within critical scholarship on the commons, more generally. Indeed, this book demonstrated how capital incorporates FLOSS production into commercial offerings in various ways. To actively promote the growth of both the subjective and objective qualities of the commons, commons-based movements will actively need to work to subvert capital logics by positioning their activities in an antagonistic relationship to capital.

By seeking reformist agendas from within existing institutions, such movements risk remaining small-scale, fragmented, and only capable of temporary subsistence rather than formulating a coordinated alternative to prevailing logics. Therefore, commons-based movements need to move beyond a *politics of provision* (based on the granting of individual rights, open access, etc.). Such a politics would not only provide rights of access to community members, but the sources of their commonwealth would also continue to be susceptible to capital and state appropriation. To be sure, the inroads made by movements informed by liberal–democratic political economy have led to the widespread adoption of particular commons-based resources (see especially Linux and the technologies of free and open source software). But insofar as these resources are available to capital, they only exacerbate or accelerate the inequities involved in circuits of capital accumulation.

One of the most well-developed proposals for reforming existing institutions to bring about a commons-based society comes from the P2P Foundation (2019) and its Commons Transition Plan. The plan outlines policy prescriptions toward a commons-based society where citizens are treated as commoners. As I have outlined throughout this paper, however, the dilemma of how to ensure that the value created by commons-based movements remains within the commons persists. Bauwens and Niaros (2017) explore this dilemma through an analysis of value within the commons economy. The authors argue that economic theory is experiencing a 'value crisis' in light of the emergent practices of commons-based communities. They argue that whereas value within capitalism is *extractive*, a shift to a *generative* value model would enrich the communities and resources directly involved in production. The open cooperative and platform cooperative (Scholz, 2014) are organisational forms that have been developed as a means for directly enriching those involved in production. However, the specific tactics used by open cooperatives to ensure that the value created by their contributors stays within the commons varies. Bauwens and Niaros (2017) provide case studies that illustrate these differences. Most important for the purpose of my argument, however, is the question of how value can be actively re-appropriated from capital and placed into the commons value circuit.

My argument is that we need a form of 'subversive commoning', which would actively seek to incorporate resources into commons value circuits. Just as capital operates according to a logic of capital accumulation by dispossession (Harvey, 2004), so too can commons-based movements reverse this logic to establish a site of social struggle. This could be framed as *commons pooling by capital dispossession*, although there are a couple of caveats to such an expression. First, I use the term 'pooling' here to signal an opposition to the private accumulation of capital. However, commons-based movements need to find ways of actively growing their commoning capacity over time. Doing so could accelerate the pace of the social revolution described by Marx, as well as more recently by De Angelis. Second, 'dispossession' is not necessarily an entirely accurate term when applied to the digital commons. Rather, digital resources could be appropriated by commons-based movements to serve their own needs.

Bauwens and Niaros (2017) use the term 'reverse co-optation' to describe the ways in which commons-based movements can 'use capital from the capitalist or state system, and subsume capital to the new logic' of the commons (3). The example given by the authors is the open cooperative, Enspiral, which uses a policy of 'capped returns' to protect its operations from the perpetual returns that investors often seek when investing in a company. In essence, shares in a new company are offered to investors along with an option for the company to repurchase those shares at an agreed upon price in the future. The idea is that the interests of the investor and the cooperative become aligned; both have an interest in seeing the cooperative succeed. The investor will be guaranteed some return on the initial investment, and the cooperative will have full control of its finances. In the case of Enspiral, once the capped return contract has been fulfilled, all resources are then given to the commons. In this sense, Enspiral provides an example of how an open cooperative can actively grow common-pool resources.

While Enspiral provides one example of how the commons can grow, my idea for 'subversive commoning' would include many other examples. At a general level, we can think of movements to reclaim farming, housing, forests, and other natural resources by either occupying abandoned space or actively resisting the enclosure of ancestral lands. These activities are directly subversive to capital because they actively re-appropriate sites of capitalist production into cooperative or commons-based movements. But we also have examples from within the digital commons. For example, organisations like RiseUp or Saravá provide 'online communication tools for people and groups working on liberatory social change' (RiseUp, 2019). In addition, FemHack provides a space for feminist and queer hackers to 'hack patriarchy, capitalism, and other systems of oppression', and the group actively works to encode non-hierarchical values into their technologies and networked infrastructures (foufem, 2016). These organisations, which have been effectively built from nothing, have the subversion of the logic of capital at the core of their foundational principles. Apart from within organisations that provide digital infrastructures, tools, and services to assist in the project of bringing about social change, subversive commoning can also be seen in attempts to release knowledge and information that has been closed off from public access. Aaron Schwartz's downloading and release of academic articles held in the JSTOR database provides an example of commoning knowledge that was enclosed by the capitalist logic of publishing companies. What all these examples have in common is the subversive nature of their activities in attempting to undermine prevailing capitalist logics that either enclose knowledge and information behind paywalls or institute hierarchical systems of management, surveillance, and control over information resources. Any attempt to subvert these logics could provide an example of subversive commoning. Subversive commoning responds by appropriating these resources and re-encoding them within the logics of commons value circuits as well as within subjectivities that emphasise care, trust, mutual aid, and conviviality, while recognising the social value in social production.

By incorporating a critique of capitalism within commons-based movements, we can move closer to truly anti-capitalist commons. Caffentzis and Federici (2014) describe anti-capitalist commons in the following way:

> Anti-capitalist commons, then, should be conceived as both autonomous spaces from which to reclaim control over the conditions of our reproduction, and as bases from which to counter the processes of

enclosure and increasingly disentangle our lives from the market and the state. Thus they differ from those advocated by the Ostrom School, where commons are imagined in a relation of coexistence with the public and with the private. Ideally, they embody the vision that Marxists and anarchists have aspired to but failed to realize: that of a society made of 'free associations of producers', self-governed and organized to ensure not an abstract equality but the satisfaction of people's needs and desires. (Caffentzis & Federici, 2014: 101)

Rigi's (2014) proposals for 'revolutionary peer producing cooperatives' have some of these hallmarks as well. His criteria for such cooperatives are two-fold: 1) 'the cooperatives must be revolutionary', and 2) 'they must break with the market as much as they can' (401). In visualising how material production would pair with knowledge commons, Rigi claims that each cooperative would produce its own food on its commons of land, but the material commons (land, food, etc.) would only belong to the members rather than be open for all like the knowledge commons. He also claims that the cooperative must be open to new members, but there would be a cap on the total number of people who are allowed to join, which would be determined by the number of people who can be supported by the land. Rigi also suggests that any surplus of material goods could be made available to other cooperatives through a networked system of exchange between other revolutionary cooperatives. Therefore, these communities should try to develop their own communication and transportation networks to the greatest extent possible. To reduce the distances between such communities, Rigi envisions such cooperatives to be a series of smaller communities (approximately 200,000), which would require massive movements of people out of urban centers and back to the countryside. The goal here is to reduce the strain on urban environments and ecologies, while revitalising some of the areas that have been left behind as now more than half the world's population resides in urban areas.

Undoubtedly, there will be disagreements on how to most effectively accomplish such a mass mobilisation. The end goal, however, is to design a more equitable and sustainable future for the planet and people. While this may seem like an unobjectionable goal, too often progressive social movements become mired in debates about the appropriate means to achieve these goals, as if there were one singular means for achieving social change. My own view is that we ought not to be entirely dismissive of any effort at bringing about change, especially if that change is aimed at combatting the injustices of global capitalism. Rather, to truly mount a substantive challenge to the tendencies of global capital, we will require a multifaceted approach that accounts for the unique specificities within local contexts. The point is not to provide a general prescription for how things ought to be done. Rather, as Marx reminds us, the point is to change the world. And change requires that we remain open to the unique histories, challenges, and opportunities with which we are presented.

6.7. Concluding Thoughts on Capital and the Commons

As the quote from Raymond Williams at the beginning of this chapter reminds us, technologies are just one part of a more general social struggle. Commons-based peer production, such as the type occurring within FLOSS communities, should not be viewed as a comprehensive solution to the unequal social relations of a capitalist system. Rather, commons-based peer production may be viewed as one part of a broader social struggle against global capital. More specifically, commons-based peer production can be viewed within the context of a broader resistance movement that seeks to reclaim commons of all types, whether they be tangible goods like land, water, and air, or the intangible goods of data, information, or knowledge that provide the infrastructure for social relations.

When Karl Polanyi authored *The Great Transformation*, he critiqued the then-emerging market fundamentalism of the Austrian School of economics, exemplified by Friedrich Hayek and inspired by the work of Ludwig von Mises, for its dis-embedding of market relations from social relations. For Polanyi, the market and market relations had historically been embedded within social relations, such that the social bonds connecting communities of people together were not subjected to a market logic. Rather, the market existed within and as a part of social relations. This, however, transformed after the market became elevated to a degree whereby all other relations became moulded according to its logic. This dis-embedding of the market from social relations has the normative effect of creating certain 'fictitious commodities,' like land, labour, and money that had all previously been important infrastructural elements of social life. In other words, when land becomes a commodity, concerns about its long-term sustainability become subsumed under a market logic that seeks profit from its exploitation. The same applies to labour, which is to say, human beings, who become exploited and valued according to a market logic. Finally, money becomes something to be hoarded for its intrinsic or future value rather than its function as a universal equivalent for exchanging different goods.

Polanyi's critique could, perhaps, be expanded to include information as a fictitious commodity. This would offer a framework for situating information dialectically between the market and social relations, as well as the increasing tendency to extract information out of its social function and treat it as a commodity. Indeed, Schiller (2007) draws this distinction between information as a commodity and information as a resource. When treated as a commodity and enclosed by intellectual property protections, information becomes highly valued as a privileged resource that can only be accessed by those who are willing to pay for access. When treated as a resource and made freely available for all, information can be studied, modified, adapted, and redistributed to others who can also benefit from access to it. Thus, we arrive at two conceptualisations of information: as a privatised resource transformed into a commodity, and as a commonly held resource available for all.

Corporations, like Microsoft, have sought to transform information into a privatised resource that can be protected by copyright. The FLOSS community has sought ways to preserve information as a commonly held resource for all to use, most notably through copyleft licences like the GPL. By doing so, the community has been able to establish a knowledge commons that resists enclosure. However, the knowledge commons under capitalism may be facing a similar fate to the commons of the past, although with certain careful distinctions. This project has demonstrated how capital has readjusted its relatively inflexible position in relation to commons-based production. It needed to reorient its strategies to incorporate without enclosing the commons. By doing so, capitalist firms pursue profits while finding a variety of ways to give back to the community, whether by making code freely available under free software or open source licensing, or by supporting the informal institutions that govern various open source projects. While this may provide ad hoc support for commons-based production, it may not provide a long-term solution to commons-based labour. Instead, commons-based peer labour may be placed in an ever-more-precarious position of depressed or non-existent wages while corporations make commercial use of their contributions. What will be needed as this type of involvement continues is a sustainable way to protect the commons, but also a way to ensure investment in commons-based peer labour. In other words, not just investment in institutions, organisations, technologies, or innovations, but long-term and sustainable investment in the true source of their value, which is to say, people.

Notes

[36] The ideas in this section originally appeared in Benjamin Birkinbine. 2018. Commons Praxis: Toward a Critical Political Economy of the Digital Commons. *tripleC* 16(1): 290–305.

[37] For more on these contradictions and a critical call for media and communication scholars to formulate a newly emergent politics of the left, see Fenton, Natalie. 2016. *Digital, Political, Radical*. Cambridge, UK: Polity.

References

Amadeo, Ron. 2018, 21 July. Google's Iron Grip on Android: Controlling Open Source By Any Means Necessary. *Ars Technica*. Last accessed 4 January 2019 from http://arstechnica.com/gadgets/2018/07/googles-iron-grip-on-android-controlling-open-source-by-any-means-necessary/

Andrejevic, Mark. 2007. *ISpy: Surveillance and Power in the Interactive Era*. Lawrence, KS: University Press of Kansas.

Andrejevic, Mark. 2012. Exploitation in the Data Mine. In *Internet and Surveillance: The Challenges of Web 2.0 and Social Media*, edited by Christian Fuchs, Kees Boersma, Anders Albrechtslund and Marisol Sandoval, 71–88. New York, NY: Routledge.

Apple Computer, Inc. v. Microsoft Corporation, 35 F.3d 1435 (1994).

Bagdikian, Ben. 2004. *The New Media Monopoly*. Boston, MA: Beacon Press.

Barbrook, Richard and Andy Cameron. 1995. The Californian Ideology. *Mute*, 1(3). Last accessed 5 December 2018 from http://www.metamute.org/editorial/articles/californian-ideology

Bauwens, Michel. 2005. The Political Economy of Peer Production. *CTheory*. Last accessed 10 September 2019 from http://ctheory.net/ctheory_wp/the-political-economy-of-peer-production/

Bauwens, Michel. 2013. Thesis on Digital Labour in an Emerging P2P Economy. In *Digital Labour: The Internet as Playground and Factory*, edited by Trebor Scholz, 211–224. New York, NY: Routledge.

Bauwens, Michel and Vasilis Kostakis. 2014. From the Communism of Capital to Capital for the Commons: Toward Open Co-Operativism. *TripleC* 12(1): 356–361.

Bauwens, Michel and Vasilis Niaros. 2017. Value in the Commons Economy: Developments in Open and Contributory Value Accounting. Report published by the P2P Foundation. Last accessed 4 January 2019 from https://www.boell.de/sites/default/files/value_in_the_commons_economy.pdf

BBC News. 2004, 26 August. Microsoft Linux Ad 'Misleading.' *BBC News*. Last accessed 4 December 2018 from http://news.bbc.co.uk/2/hi/technology/3600724.stm

Benkler, Yochai. 2006. *The Wealth of Networks: How Social Production Transforms Markets and Freedom*. New Haven, CT: Yale University Press.

Berners-Lee, Tim, and Robert Caillau. 1990, November 12. WorldWideWeb: Proposal for a Hypertext Project. Last accessed 6 December 2018 from: http://www.w3.org/Proposal.html

Bettig, Ronald. 1992. Critical Perspectives on the History and Philosophy of Copyright. *Critical Studies in Mass Communication* 9(2): 131–155.

Birkinbine, Benjamin. 2016a. Free Software as Public Service: An Assessment of Activism, Policy, and Technology. *International Journal of Communication* 10: 3893–3908. Available via open access from https://ijoc.org/index.php/ijoc/article/view/4974

Birkinbine, Benjamin. 2016b. Conflict in the Commons: Toward a Political Economy of Corporate Involvement in Free and Open Source Software. *The Political Economy of Communication* 2(2): 3–19.

Birkinbine, Benjamin. 2017. From the Commons to Capital: Red Hat, Inc. and the Business of Free Software. *Journal of Peer Production* 10.

Birkinbine, Benjamin. 2018. Commons Praxis: Toward a Critical Political Economy of the Digital Commons. *TripleC* 16(1): 290–305.

Birkinbine, Benjamin, Rodrigo Gómez and Janet Wasko, eds. 2017. *Global Media Giants*. New York, NY: Routledge.

Bloomberg. 2013, 20 January. Oracle Faces In-Depth EU Probe Over Sun Purchase. *Business Standard*. Last accessed 2 January 2019 from https://www.business-standard.com/article/companies/oracle-faces-in-depth-eu-probe-over-sun-purchase-109090400042_1.html

Boltanski, Luc and Eve Chiapello. 2005. *The New Spirit of Capitalism*. London: Verso.

Boyle, James. 2003. The Second Enclosure Movement and the Construction of the Public Domain. *Law and Contemporary Problems* 66: 33–74. Last accessed May 26, 2014 from http://scholarship.law.duke.edu/cgi/viewcontent.cgi?article=1273&context=lcp

Braverman, Harry. 1974. *Labor and Monopoly Capital: The Degradation of Work in the Twentieth Century*. New York, NY: Monthly Review Press.

Bridgewater, Adrian. 2013, 13 May. International Space Station Adopts Debian Linux, Drops Windows & Red Hat Into an Airlock. *Computer Weekly*. Last

accessed September 18, 2018 from https://www.computerweekly.com/blog/
Open-Source-Insider/International-Space-Station-adopts-Debian-Linux-
drops-Windows-Red-Hat-into-airlock

Broumas, Antonios. 2017a. The Ontology of Intellectual Commons. *International Journal of Communication* 11: 1507–1527. Last accessed 24 June 2019
from https://ijoc.org/index.php/ijoc/article/view/6347

Broumas, Antonios. 2017b. Social Democratic and Critical Theories of the
Intellectual Commons: A Critical Analysis. *TripleC* 15 (1), 100–126.

Caffentzis, George and Silvia Federici. 2014. Commons Against and Beyond
Capitalism. *Community Development Journal* 49(1): 92–105.

Campbell-Kelly, Martin. 2001. Not Only Microsoft: The Maturing of the Personal Computer Software Industry, 1982–1995. *The Business History Review*
75(1): 103–145. Last accessed 4 December 2018 from http://www.jstor.org/
stable/3116558

Chan, Sharon Pian. 2011, 11 May. Long Antitrust Saga Ends for Microsoft. *The
Seattle Times*. Last accessed 4 December 2018 from http://seattletimes.com/
html/microsoft/2015029604_microsoft12.html

Coleman, E. Gabriella. 2004. The Political Agnosticism of Free and Open
Source Software and the Inadvertent Politics of Contrast. *Anthropological
Quarterly* 77(3): 507–519.

Coleman, E. Gabriella. 2013. *Coding Freedom: The Ethics and Aesthetics of
Hacking*. Princeton, NJ: Princeton University Press.

Copeland, B. Jack. 2006. The Modern History of Computing. *Stanford Encylo-
pedia of Philosophy*. Last accessed 5 January 2019 from http://plato.stanford.
edu/entries/computing-history/

Corbet, Jonathan & Greg Kroah-Hartman. 2017. *Linux Kernel Development
Report*. The Linux Foundation. Last accessed 13 September 2018 from
https://www.linuxfoundation.org/2017-linux-kernel-report-landing-page/

Cox, Nicole and Silvia Federici. 1976. *Counter-Planning from the Kitchen:
Wages for Housework: A Perspective on Capital and the Left*. New York, NY:
New York Wages for Housework Committee.

Cusumano, Michael A. and David B. Yoffie. 1998. *Competing on Internet Time: Lessons from Netscape and its Battle with Microsoft*. New York, NY: The Free Press.

Dalla Costa, Mariarosa and Selma James. 1975. *The Power of Women and the
Subversion of the Community*. Bristol: Falling Wall Press.

Dardot, Pierre and Christian Laval. 2019. *Common: On Revolution in the 21st
Century*. London: Bloomsbury Academic.

DeAngelis, Massimo. 2017. *Omnia Sunt Communia: On the Commons and the
Transformation to Postcapitalism*. London: Zed Books.

Deek, Fadi P. and James A. M. McHugh. 2008. *Open Source: Technology and
Policy*. New York, NY: Cambridge University Press.

Deleris, Bertrand. 2006, 1 December. Battling Bugs: Embedded Debugging Tactics. *EDN*. Last accessed 5 January 2019 from http://edn.com/
electronicsnews/4317260/Battling-bugs-embedded-debugging-tactics

Driscoll, Kevin. 2015. Professional Work for Nothing: Software Commerciali-
zation and 'An Open Letter to Hobbyists'. *Information and Culture: A Journal of History* 50(2): 257–283.

Dulong de Rosnay, Melanie and Francesca Musiani. 2016. Towards a (De)Cen-
tralization-Based Typology of Peer Production. *TripleC* 14(1): 189–207.

Dyer-Witheford, Nick. 2006. The Circulation of the Common. Paper presented
at Immaterial Labour, Multitudes and New Social Subjects: Class Composi-
tion in Cognitive Capitalism conference. University of Cambridge.

Elstrom, Peter. 1997, 22 January. Microsoft's $8 million Goodbye to Spyglass.
Businessweek.com. Archived version last accessed 4 December 2018 from
https://www.landley.net/history/mirror/ms/new0122d.htm

Federici, Silvia. 2012. *Revolution at Point Zero: Housework, Reproduction, and
Feminist Struggle*. Oakland, CA: PM Press.

The Fedora Project, 2019. Fedora Council Charter. Last accessed 10 September
2019 from https://docs.fedoraproject.org/en-US/council/

Fenton, Natalie. 2016. *Digital, Political, Radical*. Cambridge, UK: Polity.

Festa, Paul. 2001. Governments Push Open-Source Software. *CNET News*. Last
accessed 5 January 2019 from http://news.cnet.com/2100-1001_3-272299.html

Fitzpatrick, Alex. 2012, 10 August. What is the Syrian Electronic Army?
Mashable.com. Last accessed 5 January 2019 from http://mashable.
com/2012/08/10/syrian-electronic-army/

Fogel, Karl. 2005. *Producing Open Source Software: How to Run a Successful Free
Software Project*. Sebastopol, CA: O'Reilly Media.

Foley, Mary Jo. 2015, 17 April. Microsoft Shutters its Standalone Open Tech
Open-Source Subsidiary. *ZDNet*. Last accessed 13 December 2018 from
https://www.zdnet.com/article/microsoft-shutters-its-standalone-open-
tech-open-source-subsidiary/

foufem. 2019. FemHack! Last accessed 4 January 2019 from http://foufem.wiki.
orangeseeds.org/

Free Software Foundation, Inc. 2012. The Free Software Definition. Last accessed
5 January 2019 from https://www.gnu.org/philosophy/free-sw.html

Frischmann, Brett. 2012. *Infrastructure: The Social Value of Shared Resources*.
Oxford: Oxford University Press.

Fuchs, Christian. 2008. *Internet and Society: Social Theory in the Information
Age*. New York, NY: Routledge.

Fuchs, Christian. 2011a. Web 2.0, Prosumption, and Surveillance. *Surveillance &
Society* 8(3): 288–309. Last accessed 20 December 2018 from https://ojs.
library.queensu.ca/index.php/surveillance-and-society/article/view/4165

Fuchs, Christian. 2011b, 1 February. New Media, Web 2.0 and Surveillance.
Sociology Compass 5(2): 134–147.

Fuchs, Christian. 2012. Critique of the Political Economy of Web 2.0 Surveillance. In *Internet and Surveillance: The Challenges of Web 2.0 and Social
Media*, edited by Christian Fuchs, Kees Boersma, Anders Albrechtslund
and Marisol Sandoval, 31–70. New York, NY: Routledge.

Fuchs, Christian. 2013. Class and Exploitation on the Internet. In *Internet and Surveillance: The Challenges of Web 2.0 and Social Media*, edited by Christian Fuchs, Kees Boersma, Anders Albrechtslund and Marisol Sandoval, 211–224. New York, NY: Routledge.

Fuchs, Christian. 2015. *Digital Labour and Karl Marx*. London: Routledge.

Fydorenchyk, Tetiana. 2014, 6 February. Software Stacks Market Share: January 2014. *Jelastic*. Last accessed 3 January 2019 from http://blog.jelastic.com/2014/02/06/software-stacks-market-share-january-2014/

Gallagher, Sean. 2013, 18 October. The Navy's Newest Warship is Powered by Linux. *Ars Technica*. Last accessed on 18 September 2018 from https://arstechnica.com/information-technology/2013/10/the-navys-newest-warship-is-powered-by-linux/

Garland, Harry. 1977. Design Innovations in Personal Computers. *Computer* 10(3): 24–27. DOI:10.1109/C-M.1977.217669

Gates, Bill. 1976. An Open Letter to Hobbyists. In *Homebrew Computer Club Newsletter*, 2(1), edited by Robert Reiling. Mountain View, CA: Hombrew Computer Club. Available from Digibarn Computer Museum, last accessed 4 December 2018 from http://www.digibarn.com/collections/newsletters/homebrew/V2_01/index.html

Gilbert, Richard J. 1995. Networks, Standards, and the Use of Market Dominance: Microsoft. In *The Antitrust Revolution: Economics, Competition, and Policy*, edited by John E. Kwoka Jr. and Lawrence J. White. 2004. New York, NY: Oxford University Press, 409–429.

Gleick, James. 2011. *The Information: A History, A Theory, A Flood*. New York, NY: Pantheon Books.

Gramsci, Antonio. 1971. *Selections from the Prison Notebooks*. New York, NY: International Publishers.

Greene, Thomas C. 2001. Ballmer: 'Linux is a Cancer.' *The Register*. Last accessed 5 January 2019 from http://www.theregister.co.uk/2001/06/02/ballmer_linux_is_a_cancer/

Gutierrez-Aguilar, Raquel. 2014. Beyond the 'Capacity to Veto': Reflections from Latin America on the Production and Reproduction of the Common. *South Atlantic Quarterly*, 113(2): 259–270.

Hamm, Steve and Jay Greene. 2004. The man who could have been Bill Gates. *Bloomberg Businessweek Magazine*. Last accessed 5 December 2018 from http://www.businessweek.com/stories/2004-10-24/the-man-who-could-have-been-bill-gates

Hardin, Garrett. 1968. The Tragedy of the Commons. *Science* 162 (3859): 1243–1248.

Hardt, Michael and Antonio Negri. 2004. *Multitude: War and Democracy in the Age of Empire*. New York, NY: Penguin Books.

Hardt, Michael and Antonio Negri. 2009. *Commonwealth*. Cambridge, MA: Belknap Press of Harvard University Press.

Harmon, Amy and John Markoff. 1998, 3 November. Internal memo shows Microsoft executives' concern over free software. *The New York Times*.

Archived version last accessed 4 December 2018 from http://www.nytimes.com/library/tech/98/11/biztech/articles/03memo.html

Harvey, David. 1989. *The Condition of Postmodernity: An Enquiry Into the Origins of Cultural Change.* Oxford: Blackwell.

Harvey, David. 2004. The 'New' Imperialism: Accumulation by Dispossession. *Socialist Register* 40: 63–87.

Hayes, Mike. 1976, 20 February. 'Regarding your letter of 3 February 1976 appearing in Homebrew Computer Club Newsletter vol. 2 no. 1'. In *Homebrew Computer Club Newsletter*, 2(2). Mountain View, CA: Homebrew Computer Club. Available from Digibarn Computer Museum, last accessed 5 December 2018 from http://www.digibarn.com/collections/newsletters/homebrew/V2_02/homebrew_V2_02_p2.jpg

Heath, Nick. 2017, 23 November. From Linux to Windows 10: Why did Munich Switch and Why Does It Matter? *TechRepublic.* Last accessed 5 January 2019 from https://www.techrepublic.com/article/linux-to-windows-10-why-did-munich-switch-and-why-does-it-matter/

Hess, Charlotte and Elinor Ostrom. 2007. Introduction: An Overview of the Knowledge Commons. In *Understanding Knowledge as a Commons: From Theory to Practice*, edited by Charlotte Hess & Elinor Ostrom, 3–26. Cambridge, MA: MIT Press.

History of Computing Project, The. 2014. Microsoft Company 15 September 1975. Last accessed 5 December 2018 from http://www.thocp.net/companies/microsoft/microsoft_company.htm

Hoffa, Felipe. 2017, 24 June. Who Contributed the Most to Open Source in 2017? Let's Analyze GitHub's Data and Find Out. *Medium.* Last accessed 13 December 2018 from https://medium.freecodecamp.org/the-top-contributors-to-github-2017-be98ab854e87

Jarrett, Kylie. 2016. *Feminism, Labour, and Digital Media: The Digital Housewife.* New York, NY: Routledge.

Kanellos, Michael. 1999, 8 September. Red Hat Stock Surge Creates Billionaires. *CNet.* Last accessed 21 December 2018 from http://news.cnet.com/Red-Hat-stock-surge-creates-billionaires/2100-1001_3-205557.html

Kaste, Martin. 2004, 15 September. Brazil Switches from Microsoft to 'Open Source' software. National Public Radio. Last accessed 4 January 2019 from http://www.npr.org/templates/story/story.php?storyId=3919175

Kelty, Christopher M. 2008. *Two Bits: The Cultural Significance of Free Software.* Durham, NC: Duke University Press.

Kingstone, Steve. 2005. Brazil Adopts Open Source Software. *BBC News.* Last accessed 5 January 2019 from http://news.bbc.co.uk/2/hi/4602325.stm

Kleiner, Dmitry. 2010. *The Telekommunist Manifesto.* Amsterdam: Institute of Networked Cultures. Last accessed 24 June 2019 from http://media.telekommunisten.net/manifesto.pdf

Lai, Eric. 2007, 29 October. Microsoft and open-source backers eye each other – warily. *ComputerWorld.* Last accessed 13 December 2018 from https://www.

computerworld.com/article/2552248/enterprise-applications/microsoft-and-open-source-backers-eye-each-other-warily.html

Laishram, Ricky. 2010, 14 August. Oracle has Killed OpenSolaris. *Techie Buzz*. Last accessed 3 January 2019 from http://techie-buzz.com/foss/oracle-has-killed opensolaris.html

Lakhani, Karim R. and Robert G. Wolf. 2005. Why Hackers Do What They Do: Understanding Motivation and Effort in Free/Open Source Software Projects. In *Perspectives on Free and Open Source Software*, edited by Joseph Feller, Brian Fitzgerald, Scott A. Hissam, and Karim R. Lakhani, 3–21. Cambridge, MA: MIT Press.

Lazzarato, Maurizio. 1996. Immaterial Labour. In *Radical Thought in Italy: A Potential Politics*, edited by Paul Virno and Michael Hardt. Minneapolis, MN: University of Minnesota Press.

Lessig, Lawrence. 2001. *The Future of Ideas: The Fate of the Commons in a Connected World*. New York, NY: Random House.

Lessig, Lawrence. 2006. *Code: Version 2.0*. New York, NY: Basic Books.

Levy, Steven. 1984. *Hackers: Heroes of the Computer Revolution*. Garden City, NY: Anchor Press/Doubleday.

Linebaugh, Peter. 2008. *The Magna Carta Manifesto*. Berkeley, CA: University of California Press.

Linux Foundation, The. 2012. *Linux Kernel Development: How Fast it is Going, Who is Doing It, What They are Doing, and Who is Sponsoring It?* Available at https://go.linuxfoundation.org/who-writes-linux-2012

Linux Foundation, The. 2016, 16 November. Microsoft Fortifies Commitment to Open Source, Becomes Linux Foundation Platinum Member. Press Release. Last accessed 13 December 2018 from https://www.linuxfoundation.org/press-release/2016/11/microsoft-fortifies-commitment-to-open-source-becomes-linux-foundation-platinum-member/

Locher, Fabien. 2016. Third World Pastures: The Historical Roots of the Commons Paradigm, 1965–1990. *Quaderni Storici* 51(1): 303–333.

Luhmann, Niklas. 1995. *Social Systems*. Stanford: Stanford University Press.

Machlup, Fritz. 1962. *The Production and Distribution of Knowledge in the United States*. Princeton, NJ: Princeton University Press.

MariaDB Foundation. 2018. Sponsor. Last accessed 2 January 2019 from https://mariadb.org/donate/

Marx, Karl. 1864. Results of the Direct Production Process. In *Economic Works of Karl Marx, 1861–1864*. Available from The Marxist Internet Archive, last accessed 24 June 2019 from http://www.marxists.org/archive/marx/works/1864/economic/

Marx, Karl. 1906. *Capital, A Critique of Political Economy, Vol. 1*. New York, NY: Modern Library.

Marx, Karl. 1998. *The German Ideology*. Amherst, NY: Prometheus Books.

Maturana, Humberto and Francisco Varela. 1998. *The Tree of Knowledge: The Biological Roots of Human Understanding*. Boston: Shambhala.

Maxwell, Richard. 2003. *Herbert Schiller*. Lanham, MD: Rowman & Littlefield.

McGuigan, Lee and Vincent Manzerolle. 2013. *The Audience Commodity in the Digital Age: Revisiting a Critical Theory of Commercial Media*. New York, NY: Peter Lang Publishing.

McKercher, Catherine and Vincent Mosco, eds. 2007. *Knowledge Workers in the Information Society*. Lanham, MD: Lexington Books.

Meehan, Eileen R. 1999. Commodity, Culture, Common Sense: Media Research and Paradigm Dialogue. *Journal of Media Economics* 12(2): 149–163.

Meehan, Eileen R. 2005. *Why TV is Not our Fault: Television Programming, Viewers, and Who's Really in Control*. Lanham, MD: Rowman & Littlefield.

Meehan, Eileen R., Vincent Mosco and Janet Wasko. 1993. Rethinking Political Economy: Continuity and Change. *Journal of Communication* 43(4): 347–358.

Meretz, Stefan. 2014. Socialist Licenses? A Rejoinder to Michel Bawens and Vasilis Kostakis. *TripleC* 12(1): 362–365.

Microsoft. 2018, 4 June. Microsoft to Acquire GitHub for $7.5 billion. Microsoft News Center. Last accessed 13 December 2018 from https://news.microsoft.com/2018/06/04/microsoft-to-acquire-github-for-7-5-billion/

Mizokami, Kyle. 2017, 17 August. Inside the Stealth Destroyer USS Zumwalt, the Warship that Runs on Linux. *Popular Mechanics*. Last accessed 18 September 2018 from https://www.popularmechanics.com/military/weapons/news/a27804/stealth-destroyer-uss-zumwalt-linux/

Moglen, Eben. 2003. The dotCommunist Manifesto. Last accessed 24 June 2019 from http://moglen.law.columbia.edu/publications/dcm.html

Moody, Glyn. 2001. *Rebel Code: The Inside Story of Linux and the Open Source Revolution*. Cambridge, MA: Perseus Publishing.

Mosco, Vincent. 2006. Knowledge and Media Workers in the Global Economy: Antimonies of Outsourcing. *Social Identities* 12(6): 771–790.

Mosco, Vincent. 2009. *The Political Economy of Communication*. London: SAGE.

Mosco, Vincent. 2014. *To the Cloud: Big Data in a Turbulent World*. Boulder, CO: Paradigm Publishers.

Mundie, Craig. 2001, 3 May. Speech transcript – Craig Mundie, The New York University Stern School of Business. Microsoft.com. Last accessed 13 December 2018 from https://news.microsoft.com/speeches/speech-transcript-craig-mundie-the-new-york-university-stern-school-of-business/

Nairn, Alasdair. 2002. *Engines that Move Markets: Technology Investing from Railroads to the Internet and Beyond*. New York, NY: John Wiley & Sons.

Neeson, J. M. 1993. *Commoners: Common Right, Enclosure and Social Change in England, 1700–1820*. New York, NY: Cambridge University Press.

O'Mahony, Siobhán and Beth A. Bechky. 2008. Boundary Organizations: Enabling Collaboration Among Unexpected Allies. *Administrative Science Quarterly* 53(3): 422–459.

OpenOffice Community Council. 2010, 14 October. Community Council Log 201014. Last accessed 2 January 2019 from http://wiki.openoffice.org/wiki/Community_Council_Log_20101014

Open Source Ecology. 2019. Web site. Last accessed 5 January 2019 from http://opensourceecology.org/

Open Source Initiative. 2007, 12 October. OSI Approves Microsoft License Submissions. Open Source Initiative. Last accessed 13 December 2018 from http://opensource.org/node/207

O'Reilly, Tim. 2005. What is Web 2.0: Design patterns and business models for the next generation of software. *O'Reilly.com.* Last accessed 20 December 2018 from http://oreilly.com/pub/a/web2/archive/what-is-web-20.html?page=all

Ostrom, Elinor. 1990. *Governing the Commons: The Evolution of Institutions for Collective Action.* Cambridge: Cambridge University Press.

Ostrom, Elinor. 2005. *Understanding Institutional Diversity.* Princeton: Princeton University Press.

P2P Foundation. 2019. The Commons Transition Primer. Accessed 4 January 2019 from http://primer.commonstransition.org.

Page, William H. and John E. Lopatka. 2007. *The Microsoft Case: Antitrust, High Technology, and Consumer Welfare.* Chicago, IL: The University of Chicago Press.

Pang, Alex Soojung-Kim and Wendy Marinaccio. 2000. The Xerox PARC Visit. Included as part of Making the Macintosh: Technology and Culture in Silicon Valley, an ongoing project available online. Last accessed 4 December 2018 from http://web.stanford.edu/dept/SUL/library/mac/parc.html

PC World. 2008, 16 January. Sun to Acquire MySQL for $1 Billion. Available from https://www.pcworld.com/article/141409/article.html.

Pickard, Victor. 2006. United Yet Autonomous: Indymedia and the Struggle to Sustain a Radical Democratic Network. *Media, Culture & Society* 28(3), 315–336.

Polanyi, Karl. 2001. *The Great Transformation: The Political and Economic Origins of our Time.* Boston, MA: Beacon Press.

Pollack, Andrew. 1990, 24 March. Most of Xerox's suit against Apple barred. *The New York Times.* Last accessed 6 December 2018 from https://www.nytimes.com/1990/03/24/business/most-of-xerox-s-suit-against-apple-barred.html

Powell, Alison. 2012. Democratizing Production Through Open Source Knowledge: From Open Software to Open Hardware. *Media, Culture & Society* 34(6), 691–708.

Powell, Alison. 2016. Hacking in the Public Interest: Authority, Legitimacy, Means, and Ends. *New Media & Society* 18(4), 600–616.

Powell, Alison. 2018. Moral Orders in Contribution Cultures. *Communication, Culture & Critique* 11, 513–529.

Prakash, Abhishek. 2017, 9 March. With FOSS, Indian State of Kerala Saves $58 Million Each Year. *ItsFOSS*. Last accessed 5 January 2019 from https://itsfoss.com/open-source-kerala/

Raymond, Eric S. 1998a. Open source software: A (New?) Development Methodology. The Halloween Documents: Halloween Document I (Version 1.17). Last accessed 20 December 2018 from http://www.catb.org/esr/halloween/halloween1.html

Raymond, Eric S. 1998b. Linux OS Competitive Analysis: The Next Java VM? The Halloween Documents: Halloween Document II (Version 1.7). Last accessed 20 December 2018 from http://www.catb.org/esr/halloween/halloween2.html

Raymond, Eric S. 1998c. Microsoft's Reaction to the 'Halloween Memorandum'. The Halloween Documents: Halloween Document III (Version 1.6). Last accessed 20 December 2018 from http://www.catb.org/esr/halloween/halloween3.html

Raymond, Eric S. 2000. *The Cathedral and the Bazaar* [online version]. Last accessed 5 January 2019 from http://www.catb.org/esr/writings/cathedral-bazaar/

Raymond, Eric S. 2002a. Halloween VII: Survey Says. The Halloween Documents. Last accessed 20 December 2018 from http://www.catb.org/esr/halloween/halloween7.html

Raymond, Eric S. 2002b. Halloween VIII: Doing the Damage-Control Dance. The Halloween Documents. Last accessed 20 December 2018 from http://www.catb.org/esr/halloween/halloween8.html

Raymond, Eric S. 2004. Halloween X: Follow the Money. The Halloween Documents. Last accessed 20 December 2018 from http://www.catb.org/esr/halloween/halloween10.html

Red Hat, Inc. 2000–2018. Form 10-K. Annual report. Last accessed 2 January 2019 from https://investors.redhat.com/financial-information/sec-filings

Red Hat, Inc. 2006. Red Hat trademark guidelines. Last accessed 2 January 2019 from http://www.redhat.com/f/pdf/corp/RH-3573_284204_TM_Gd.pdf

Red Hat, Inc. 2016. Form 10-K. Annual report. Last accessed 2 January 2019 from https://investors.redhat.com/financial-information/sec-filings

Red Hat, Inc. 2018, 28 October. IBM to Acquire Red Hat, Completely Changing the Cloud Landscape and Becoming the World's #1 Hybrid Cloud Provider. Press Release. Last accessed 2 January 2019 from https://www.redhat.com/en/about/press-releases/ibm-acquire-red-hat-completely-changing-cloud-landscape-and-becoming-worlds-1-hybrid-cloud-provider

Reimer, Jeremy. 2005, 14 December. Total share: 30 years of personal computer market share figures. *ArsTechnica*. Last accessed 10 September 2019 from https://arstechnica.com/features/2005/12/total-share/

Rigi, Jakob. 2014. The Coming Revolution of Peer Production and Revolutionary Co-Operatives. A Response to Michel Bauwens, Vasilis Kostakis and Stefan Meretz. *TripleC* 12(1): 390–404.

RiseUp. 2019. Riseup.net. Last accessed 4 January 2019 from https://riseup.net/

Rossiter, Ned and Soenke Zehle. 2013. Acts of Translation: Organized Networks as Algorithmic Technologies of the Common. In *Digital Labor: The Internet as Playground and Factory*, edited by Trebor Scholz, 225–239. New York, NY: Routledge.

Ryan, Anne B. 2013. The Transformative Capacity of the Commons and Commoning. *Irish Journal of Sociology* 21(2): 90–102.

Santos, Carlos, George Kuk, Fabio Kon and John Pearson. 2013. The Attraction of Contributors in Free and Open Source Software Projects. *Journal of Strategic Information Systems* 22(1): 26–45.

Sayers, Sean. 2007. The Concept of Labor: Marx and His Critics. *Science and Society* 71(4): 431–454.

Schiller, Dan. 1999. *Digital Capitalism: Networking the Global Market System*. Cambridge, MA: MIT Press.

Schiller, Dan. 2007. *How to Think About Information*. Urbana, IL: University of Illinois Press.

Scholz, Trebor, ed. 2013. *Digital Labor: The Internet as Playground and Factory*. New York, NY: Routledge.

Scholz, Trebor. 2014. Platform Cooperativism vs. the Sharing Economy. *Medium*. Accessed 4 January 2019. https://medium.com/@trebors/platform-cooperativism-vs-the-sharing-economy-2ea737f1b5ad

Schoonmaker, Sara. 2009. Software Politics in Brazil: Toward a Political Economy of Digital Inclusion. *Information, Communication & Society* 12(4): 548–565.

Schoonmaker, Sara. 2018. *Free Software, the Internet, and Global Communities of Resistance*. New York, NY: Routledge.

Shaw, Aaron. 2011. Insurgent Expertise: The Politics of Free/Livre and Open Source Software in Brazil. *Journal of Information Technology & Politics* 8: 253–272.

Singh, Neera. 2017. Becoming a Commoner: The Commons as Sites for Affective Socio-Nature Encounters and Co-Becomings. *Ephemera: Theory & Politics in Organization* 17(4): 751–776.

Smith, Michael. 2011. *The Secrets of Station X: How the Bletchley Park Codebreakers Helped Win the War*. London: Biteback Publishers.

Smythe, Dallas W. 1981. *Dependency Road: Communications, Capitalism, Consciousness, and Canada*. Norwood, NJ: Ablex Publishing Corp.

Söderberg, Johan. 2011. *Free Software to Open Hardware: Critical Theory on the Frontiers of Hacking*. Ph.D. Thesis, Department of Sociology, University of Gothenburg. Gothenburg, Sweden.

Stallman, Richard M. 2002. *Free Software, Free Society: Selected Essays of Richard M. Stallman*. Boston, MA: GNU Press.

TechInsider.org. 2016. Joint Development Agreement Between IBM and Microsoft. Last accessed 6 December 2018 from https://tech-insider.org/personal-computers/research/acrobat/871126.pdf

Terranova, Tiziana. 2000. Free Labor: Producing Culture for the Digital Economy. *Social Text* 63, 18(2): 33–58.

Terranova, Tiziana. 2004. *Network Culture: Politics for the Information Age.* Ann Arbor, MI: Pluto Press.

Thompson, Edward P. 1966. *The Making of the English Working Class.* New York, NY: Vintage Books.

Thompson, Edward P. 1971. The Moral Economy of the English Crowd in the Eighteenth Century. *Past and Present* 50: 76–136.

Tiemann, Michael. 2007. Who is Behind 'Shared Source' Misinformation Campaign? Open Source Initiative. Last accessed 13 December 2018 from http://opensource.org/node/225

Top500.org. 2018a. Operating System Family/Linux. Top500.org. Last accessed 18 September 2018 from http://www.top500.org/statistics/details/osfam/1#.U4i_InKfoxA

Top500.org. 2018b. June 2018. Last accessed 18 September 2018 from https://www.top500.org/lists/2018/06/

Tramontano, Marcelo and Nilton Trevisan. 2003. A Dimensão Digital de Solonópole, Brasil. In *SIGraDi: Proceedings from the 7th Iberoamerican Congress of Digital Graphics*, 74–77. Rosario, Argentina. Last accessed 5 January 2019 from http://cumincades.scix.net/data/works/att/sigradi2003_060.content.pdf

Treanor, Paul. 2005. Neoliberalism: Origins, Theory, Definition. Last accessed 2 January 2019 from http://web.inter.nl.net/users/Paul.Treanor/neoliberalism.html

Tu, Janet I. 2012, 7 December. Goldman Sachs: Microsoft has Gone from 97 Percent Market Share of Compute [sic] Market to 20 Percent. *The Seattle Times*. Last accessed 6 December 2018 from http://seattletimes.com/html/microsoftpri0/2019853243_goldman_sachs_microsoft_os_has_gone_from_more_than.html

Turow, Joseph. 2013. *The Daily You: How the New Advertising Industry is Defining Your Identity and Your Worth.* New Haven, CT: Yale University Press.

United States Mission to European Union. 2009, 27 October. Oracle Concerned Over EU Investigation of Sun Merger. *WikiLeaks.org*. WikiLeaks cable: 09BRUSSELS1455. Last accessed 2 January 2019 from http://wikileaks.org/cable/2009/10/09BRUSSELS1455.html

United States vs. Microsoft. 84 F. Supp. 2D 9. 1999. Retrieved from Westlaw Campus database.

United States vs. Microsoft. 2000. Conclusions of Law. Retrieved from the United States Department of Justice. Last accessed 6 December 2018 from http://www.justice.gov/atr/cases/f218600/218633.htm

United States vs. Microsoft. 2002. Final Judgement. Retrieved from the United States Department of Justice. Last accessed 6 December 2018 from http://www.justice.gov/atr/cases/f200400/200457.htm

Vaughan-Nichols, Steven J. 2014, 28 March. Red Hat Reveals CentOS Plans. *CDNet*. Last accessed 2 January 2019 from http://www.zdnet.com/red-hat-reveals-centosplans-7000027812/

Von Hippel, Eric. 2005. *Democratizing Innovation*. Cambridge, MA: MIT Press.

Wallace, James and Jim Erickson. 1992. *Hard Drive: Bill Gates and the Making of Microsoft*. New York, NY: HarperCollins.

Ward, John. 2013. Apple Lore: The Creation of the Macintosh. *Vectronic's Apple World*. Archived version last accessed 5 December 2018 from https://web.archive.org/web/20100323110634/http://www.vectronicsappleworld.com/macintosh/creation.html

Weber, Stephen. 2004. *The Success of Open Source*. Cambridge, MA: Harvard University Press.

Whitney, Lance. 2009, 14 December. Oracle Pledges to Play Well with MySQL. *Cnet*. Last accessed 3 January 2019 from http://news.cnet.com/8301-1001_3-10414686-92.html

Wilcox, Joe. 2001, 13 March. Jackson Exits Microsoft Discrimination Case. *Cnet*. Last accessed 6 December 2018 from https://www.cnet.com/news/jackson-exits-microsoft-discrimination-case

Williams, Raymond. 1975. *Television: Technology and Cultural Form*. New York, NY: Schocken Books.

Williams, Sam. 2002. *Free as in Freedom: Richard Stallman's Crusade for Free Software*. Sebastopol, CA: O'Reilly.

Winseck, Dwayne. 2017. The Geopolitics of the Global Internet Infrastructure. *Journal of Information Policy* 7: 228–267.

Winseck, Dwayne and Robert Pike. 2007. *Communication and Empire: Media, Markets, and Globalization, 1860–1930*. Durham, NC: Duke University Press.

Young, Robert and Wendy G. Rohm 1999. *Under the Radar: How Red Hat Changed the Software Business – and Took Microsoft by Surprise*. Scottsdale, AZ: The Coriolis Group.

Zuse, Konrad. 1993. *The Computer: My Life*. New York, NY: Springer-Verlag Berlin Heidelberg.

Index

Bauwens, Michel   110–111, 115
Bechky, Beth A   45, 74
Benchmark Capital   76
Benkler, Yochai   25–27, 86, 102
Berg, Aurelia van den
  discussions of Linux   64–65
Berners-Lee, Tim   58
Black Duck Software   18
Bletchley Park   7
boundary commoning   112–113
boundary organisations   45, 74, 80,
    89, 92, 97, 107
  Fedora Project Board   80, 107
  Fedora Project Council   74, 80, 89
  Open Office Community
    Council   97
  OpenSolaris Community Advisory
    Board (CAB)   93
Boyle, James   24
Braverman, Harry   37, 45
Broumas, Antonios   23, 27–28,
    42–43
Browser Wars, the   58–61

C

Caillau, Robert   58
Canonical   11
capital accumulation circuits   4,
    16–17, 41–43, 45, 46, 112, 115
capitalism
  and the commons   23, 28, 33, 83,
    102, 103, 110–111
  and FLOSS   4, 16, 102
  and FLOSS labour   35–36,
    41–42, 45
CentOS   82, 103
circuits of social reproduction
    and capital accumulation
    circuits   41
Cisco   1
Coleman, E. Gabriella   15
Colossus, the   7
Common Business-Oriented
    Language (COBOL)   7

Common Development and
    Distribution License
    (CDDL)   93
commoning as a process (see also
    'boundary commoning')   23,
    43–46, 109
commons-based peer
    production   25–26
commons circuits of value   4,
    44–46, 113
commons paradigm   108–109
commons pooling   43, 115
commons, theories of   28
Cormier, Paul   77
critical political economy of
    communication   29–30,
    33–35

D

De Angelis, Massimo   28, 43–44,
    113
Deek, Fadi P.   17, 19
Defense Advanced Research Projects
    (DARPA)   7
development anthropology   21
Difference Engine (Babbage), the   6
digital commons   20–28, 44–47
  approaches to the commons
    27–29
  commons-based peer
    production   25–26
  critical theories of   42–46
  incorporation vs. enclosure
    23–25
  political economy of   33–47
digital enclosure   15, 23–26
Digital Research, Inc. (DRI)   54
Document Foundation, the   97
dot-com bubble, the   51, 76, 77, 105
  burst of   50, 66, 69, 70, 79, 91, 105
dotCommunist Manifesto
    (Moglen)   15
DRI website   71
Dulong de Rosnay, Mélanie   110

CPSIA information can be obtained
at www.ICGtesting.com
Printed in the USA
FSHW022154220220
67280FS

9 781912 656424